Alexey Popov

Magnetic Field of the Earth

Alexey Popov

Magnetic Field of the Earth

LAP LAMBERT Academic Publishing

Imprint

Any brand names and product names mentioned in this book are subject to trademark, brand or patent protection and are trademarks or registered trademarks of their respective holders. The use of brand names, product names, common names, trade names, product descriptions etc. even without a particular marking in this work is in no way to be construed to mean that such names may be regarded as unrestricted in respect of trademark and brand protection legislation and could thus be used by anyone.

Cover image: www.ingimage.com

Publisher:
LAP LAMBERT Academic Publishing
is a trademark of
International Book Market Service Ltd., member of OmniScriptum Publishing Group
17 Meldrum Street, Beau Bassin 71504, Mauritius

ISBN: 978-3-659-55111-6

Copyright © Alexey Popov
Copyright © 2014 International Book Market Service Ltd., member of OmniScriptum Publishing Group

THE TABLE OF CONTENTS

Introduction	3
THE MAGNETIC FIELD OF THE EARTH	5
The Drifting of the Magnetic Poles	5
The magnetic field and the drifting of the core	6
Cataclysms of the inversion of the magnetic field?	8
Short Information on History of Planet Earth	8
Short Information on the existing theories of "Geodynamo"	11
Geophysical research	15
Biot-Savart law	17
Radial currents in the core of the Earth	18
Amplification of the magnetic field by iron	19
The inversion of the magnetic field	20
MOVEMENT OF LITHOSPHERIC PLATES	21
The version of academician Korovyakov	21
Theory of tectonic catastrophes	21
Lithospheric plates and their driving forces	22
The form of the inner surface of the lithosphere	24
A specific character of Icelandic volcanoes	24
Will the "Icelandic clip" stand?	26
Mysteries of Iceland	27
Why does the seismic activity grow?	28
Real volcanoes instead of the mystical warming	29
The increase of tectonic activity	29
Brief information on the existing theory of the plate tectonics	30
Drift of the Earth core	33
Drift of the magnetic dipole of the Earth and the lithospheric plates	34
Deceleration of the Earth rotation	35

THE THERMAL ENERGY OF THE CORE	37
The global warming	37
The greenhouse effect	38
The human activity	39
Change of the solar activity	40
Other versions of the warming on the Earth	40
The Kyoto protocol	43
Criticism of the theory of the global warming	45
Climategate	46
Other aspects of climate change	47
North seaway will be free from ice	47
From the global warming to the global cold snap	48
A global cold snap is coming	48
Explosions are to blame for the global warming	50
The planetary climatic hypothesis	51
Milankovitch cycles	52
The Great glaciation	53
Brief information on the existing theory of emission of the Earth's inner heat	56
The thermal energy emitted by the Earth core at flow of the radial currents	57
Postamble	58
Epilogue	61
APPENDICES	63
Calculation of the magnetic field of the Earth (Appendix I)	63
Calculation of the force of the mechanism of the Earth core drift (Appendix 2)	65
Calculation of the change of the kinetic energy of the Earth (Appendix 3)	67
Calculation of the thermal energy emitted by the Earth core (Appendix 4)	68

INTRODUCTION

There are several reasons to explain why this book should be published.

1. At present Mass Media speak a lot about the global warming and the forthcoming global cooling on the Earth, they provide information about earthquakes, tsunami and volcano eruptions, as well as information about the fact that the magnetic pole of the Earth has started drifting and the inversion of the magnetic field of the Earth is coming. There have been offered different explanations of these global phenomena, but there is no answer to the question about the physical processes taking place in the core of the Earth and about the effect they produce on people's life on the Earth.

2. After examining the existing theory of "Geodynamo" which explains the appearance of the magnetic field of the Earth as the conversion of the kinetic energy of the liquid iron moving in the core of the Earth into the magnetic energy, the author proposed his idea, that the magnetic field of the Earth can be calculated through the electric charge in the core of the Earth, which is moved by the Earth rotation.

3. The "Geodynamo" theory doesn't give a distinct explanation of the inversion of the Earth's magnetic field, though paleomagnetic researches acknowledge the fact of the inversion of the magnetic field of the Earth.

4. Geophysics explains the heating of the core by the initial warmth of ~ 1000°C which appeared at the formation of the Earth, and the energy of the gravitational split of the primary homogeneous Earth into the iron core and the silicate mantle and crust, ~ 2500°C. It is regarded as axiom without any proof, because there are no other explanations nowadays.

5. The theory of the plate tectonics does not fully answer the question of the nature of formation of the driving mechanism of the lithospheric plates, while high powers do move plates and continents with earthquakes and tsunami resulting from this.

6. The global warmings and the global cold snaps are explained by "Milankovich cycles" and the greenhouse effect, but there are other causes of these global phenomena connected with the inversion of the magnetic field of the Earth.

7. Scientists link the slow-down of the Earth rotation by tidal forces of the Moon. But is that so?

The above-mentioned questions are raised in the scientific hypothesis "The Magnetic Field of the Earth" developed by the author on the basis of the laws of physics and electrodynamics with corresponding calculations which confirm the validity of the hypothesis.

THE MAGNETIC FIELD OF THE EARTH

> "A comprehensive theory of planetary magnetism is
> required, but there is no such a general theory so far"
> W. Hubbard "Inner Structure of Planets"
> Moscow, «Mir», 1987

Author provides from various sources different versions of physical processes which take place in the core of the Earth.

The drifting of the magnetic poles

The locations of the magnetic and geographical poles of the Earth do not coincide, these are absolutely different notions. Geomagnetic field cannot be called truly constant; it changes from time to time. Its principal function on the Earth is to protect life. Magnetic field lines create a specific shield around the Earth and protect its surface from the cosmic rays ruinous for life, and from the stream of charged particles of high energy. The monitoring of the Arctic magnetic pole condition (which moves towards the East-Siberian global magnetic anomaly through the Arctic Ocean) showed, for example, that by the beginning of 2002 the speed of the northern pole's movement increased from 10 km per year in the 70-ies to 40km per year in 2001. According to the data of the Institute of the Earth Magnetism we can observe unsteady drop of intensity of the Earth magnetic field. The acceleration of the poles' movement (3 km per year on average) and their movement in the corridors of the magnetic poles' inversion give reasons to suppose that we can see the repoling of the magnetic field of the Earth.[1]

---

[1] Serov M. "Global Warming". Moscow, «Knizhniy Klub Knigovek», 2010

The magnetic field and the drifting of the core

The science does not know the origin of the geomagnetic field of the Earth. At present, the theory of hydrodynamic dynamo (HD) serves as the ground. This idea was first proposed by an English physicist Larmor to explain the magnetism of the Sun, and later developed by Yan Frenkel and J. Braginskiy as applied to the Earth.

The mechanism of HD consists of the following. As a result of temperatures difference convective flows of substance appear in the liquid core of the Earth in the form of "eights", and slow down the rotation of the peripheral layers of the core and accelerate the movement of the inner layers. When the primary magnetic field is present, there must be self-excitation of HD in the system, and the electric current, which creates the magnetic field, must appear, and this is highly improbable. Mathematical description of HD is a three-dimensional problem. The equations to solve are nonlinear equations in partial derivatives. To put it clearly, it means that this task is practically unsolvable. Such kind mathematics can explain everything one wants, and prove that two by two is five.

Changes of the polarity of the magnetic field took periods of 150 thousand years, but there were discovered shorter periods of 50 thousand years. For a comparatively short period (1000-10000 years) the geomagnetic field of the Earth presumably almost disappeared, and then it restored again but with a different polarity. Within the last 76 million years, the magnetic poles of the Earth shifted their places 171 times. The last shift of the polarity of the magnetic field took place 12,5 thousand years ago.

The drifting of the magnetic poles has been registered since 1885. During the last 100 years the magnetic pole in the southern hemisphere has drifted 900 km and is already located in the Indian Ocean. The data about the condition of the Arctic magnetic pole (which moves towards the East-Siberian global magnetic anomaly through the Arctic Ocean) showed that its drifting was 120 km from 1993 until 1984 and more than 150 km from 1984 till 1994. It is peculiar that the data were first calculated, and then confirmed by particular measuring of the location of the North Pole.

Towards the drifting of the inner core between Australia and the Antarctic the strength of the magnetic field is growing and reaches 0,7 oersted, nearly as much as at the pole. On the opposite side of the Earth in the southern part of the Atlantic Ocean the strength of the magnetic field has vice versa diminished by 10%. In future, all the above-mentioned changes must be still increasing.

While examining the deep-water part of the Pacific Ocean with the help of hydrophones which had been taken from submarines, New Zealand and American oceanographers registered different sounds in the sonic and infrasonic wave range. These sounds are not similar to the squeak of whales or other animals and are so loud in amplitude that they cannot be produced by any living organisms. In deep-water cavities where the Earth crust is the thinnest, one can hear gurgling, hissing, crackle and long "groans". Thus, the mysterious processes going on in the centre of the Earth let us know about themselves. The drifting of the core of our planet must somehow give itself away, but due to the immense moment of momentum of the Earth as compared with the inner core, these changes will be insignificant. First of all it must effect in slowing down of the rotation of our planet. In 1991 the duration of the astronomical day increased by 1 second, in July 1992 the amendment added 1 more second to the day, and in 1993 another 2 seconds were introduced.

The Earth represents a peculiar three-degree-of-freedom gyroscope. If the movement of the inner core towards the earth surface continues, then presently the mass centre of the planet will shift so much that the Earth will simply somersault in space like a whirligig with a deposed centre of gravity, to take a more stable position of the rotation axis. The shift of the inner core of our planet can happen all of a sudden, under the influence of outer factors, i.e. at summation of lunar and solar tides[1].

---

[1] Simonov V. "Earthquakes, tsunami, catastrophes", Moscow, "Eksmo". 2011

Cataclysms of the inversion of the magnetic field?

In 1979 based on the analysis of the Earth crust samples Swedish scientists N. Merner, J. Lancer, and J. Hospers discovered that in approximately 10430 B.C. the inversion of the magnetic field of our planet took place, i.e. its polarity changed.

In the book «Space impacts and evolutions of biosphere" by B. Vladimirskiy and L. Kislovskiy the authors speak about the discovered thin layer of nitrogenous elements which are present at the bottoms of lakes of all continents. These are traces of the powerful gamma radiation, which influenced our planet in 10.000 B.C.

A trustworthy scientific magazine 'Science", January 1993, described the scale of the disastrous flood which took place 12-13 thousand years ago in Asia. A 450-meter-high wave traveling at high speed splashed out of the Caspian Sea and reached the Altai Territory destroying everything on its way and partially breaking through the mountains. American scientists found traces of two gigantic tsunami 500 and 800 meters high, which long ago came down the Pacific coast of South America. The traces of the catastrophic effects of the torrents of water can be observed long distances away from the ocean.

The Nobel Prize laureate W. Libby stated that approximately 10400 years ago traces of any activity of the primitive man disappeared on the American continent. The same decline of the humanity development can be observed in Europe, Egypt, South America and Central Asia[1].

Short Information on History of Planet Earth, evolution of plate tectonics and the modern conception of the movement of the Earth's continents

The majority of people who thoroughly examine the map of the world (usually with the Atlantic Ocean in its centre) could notice that if we remove the ocean, the cutouts of its coastlines will coincide. In spite of the fact that thousands of people

---

[1] Simonov V. "Earthquakes, tsunami, catastrophes", Moscow, "Eksmo". 2011

could have noticed this peculiarity, it was not until the beginning of the XX century that the consequences of this observation were seriously considered. It was then that Alfred Wegener, a German meteorologist, began to collect and explore information about the flora and fauna of the continents separated by the Atlantic Ocean. He also carefully studied all facts known about their geology and paleontology and about the fossil remains of organisms found there. Analyzing the findings Wegener made an inescapable conclusion that in the distant past all continents had constituted a single unit. He discovered that some outlines of the geological composition of South America which are abruptly cut by the coastline of the Atlantic Ocean, have their continuation in Africa. When having cut them from the map he joined these continents like pieces of a gigantic puzzle, the geological peculiarities of the continents coincided as if to continue each other. He also discovered that there is geological evidence of ancient glaciations which occupied India, Australia and South Africa at approximately the same period. He also found that all these continents can be joined in such a way that the areas of their glaciations would form a united territory.

In 1915 in Germany he published a book under the title "The Origin of Continents and Oceans". In this book he thoroughly analyzed this evidence and proposed his theory of "continental drift". Still in spite of the amount of the geological data, Wegener missed many important details, and was rather liberal in selecting facts to support his theory. Partially due to this fact, his hypothesis was not taken seriously that time. Moreover, prominent physicists of that period stated that outer parts of the Earth are too hard to let continents drift like ships in the sea. In particular, they indicated that the forces, which Wegener regarded as movers of the continents (the centrifugal forces appearing because of the Earth's axial rotation) are too weak for such a task. Wegener's ideas failed because of the absence of a suitable mechanism: it was presumed that without a reasonable driving force the drift of the continents is impossible.

Rehabilitation of Wegener's ideas in the form of the plate tectonics theory happened as a result of the ocean floor researches carried on in 1950-ies and 1960-ies with the help of echo sounders. The ocean floor revealed huge mountain ranges, deep canyons, immense volcanoes and steeps. Especially prominent is a mountain ridge, which stretches along the axis of the Atlantic Ocean, which is partly due to the ocean's central position on the map. The Mid-Atlantic Ridge dissects the ocean exactly in the middle copying all the dents and cavities of the coastline of its both sides, thus roughly cutting the map into two parts. It is on average 2,5 km high above the deepest parts of the ocean which are situated to the west and to the east of it. On the biggest part of its area, just along the centerline there is a rift, that is either a gap or valley with steep slopes. In the northern part of the Atlantic Ocean the Mid-Atlantic Ridge rises above the surface of the ocean forming the island of Iceland. The ridge that crosses the Atlantic is in fact a part of a continuous ridge system, which stretches through all the oceans. Investigation of the ocean floor explained the origin of the system of ocean ridges and led to the development of the plate tectonics theory from a completely unexpected source – namely from investigation of the magnetic features of the rocks of the ocean floor. When oceanological ships first began to tug magnetometers behind themselves, it was discovered that the pattern of the magnetic anomalies on the ocean floor, which depicts the magnetic features of the rocks on the ocean floor, is perfectly regular. When the locations of the measuring points and the values of the strength of the magnetic field were mapped, the isolines drawn created a zebra-looking pattern on the map of the magnetic field intensity. When igneous rock cool down from the initial molten state, some iron-bearing minerals that form from it are magnetized by the magnetic field of the Earth.

Geological and geophysical mapping along the routes transversely oriented to the stretching of the Mid-Atlantic Ridge showed that the rock, which is exactly above the axis of the ridge, is strongly magnetized in the direction of the modern magnetic field, just as it was expected. But the symmetric zebra-looking pattern of the magnetic field indicates that the sea-bed is differently magnetized in different stripes parallel to the stretch of the ridge. Some of these stripes are normally magnetized, just like the stripes located on the axis of the ridge. But they alternate with the stripes which are oppositely magnetized, as if when these stripes appeared, the north and the south magnetic poles of the Earth had changed their places.

The scientists understood, that it was not once that the magnetic field of the Earth changed its polarity during the geological time! They also understood that the magnetic stripes of the sea-bed, the 180° fluctuations of the polarity direction, and the drift of the continents are interconnected phenomena. These observations convinced the majority of the geologists that the expansion of the sea floor away from the oceanic fracture is a reality. The new oceanic crust is formed by the lava, which continuously flows from the depths in the axial parts of oceanic ridges. The magnetic pattern of the rock of the ocean floor is symmetrical on both sides of the ridge axis, because the newly coming portion of lava is magnetized at cooling into hard rock and is evenly expanding on both sides of the medial fracture. Though the speed of the formation of the ocean floor varies from place to place, its value, calculated by magnetic anomalies, usually equals several centimeters per year. The continents located on opposite sides of the Atlantic Ocean move at this speed apart from each other. Though several cm per year is really slow, the whole Atlantic Ocean at such speed could have taken 200 million years to form, which is not so much from the geological point of view. On both sides of the Atlantic Ocean the continents are firmly adhered to the rock of the ocean floor. In their movement the continents do not plough their way through the hard rock of the ocean floor; both the continents and the oceanic crust move together as a whole, being parts of a common lithospheric plate. The mechanism of the plates' movement, their actual driving force is not yet thoroughly studied.[1]

Short Information on the Existing Theories of "Geodynamo"

The essence of the dynamo mechanism is explained as transformation of the kinetic energy of the movements of current-conducting liquid into magnetic energy.

---
[1] J. MacDougall "Short History of Planet Earth". Transl. from English, St. Petersburg. "Amfora", 2001

The explanation of the origin of geomagnetic field by the dynamo mechanism is rather complicated. The point is in well-known Lenz rule, according to which the additional electric current, which appears in the current loop moving in the initial magnetic field, is directed so as to reduce the initial magnetic field, i.e. simplest flows of conductive liquid including differential rotation cannot lead to self-excitation of the magnetic field. To avoid the Lenz rule, two current loops are required, which reciprocally intensify the magnetic fields running through these loops. The idea was first outspoken by an American scientist Eugene Parker in 1955. He pointed out a concrete realistic mechanism leading to self-excitation of the magnetic field of the Sun as a result of the influence of electromagnetic induction; in the middle of 60-ies it was suggested for geodynamo by S. I. Braginskiy, and its complete formulation was offered by German scientists Steenbeck, Krause and Redler in 1984.

The essence of the idea is as follows. The magnetic field in the vacuum, which is created by the electric current, is perpendicular to this current. However, it turned out that the magnetic field in a random, turbulent or convective flow acquires a component parallel to the electric current. It is agreed to indicate the corresponding coefficient by the letter α, and the phenomenon itself got the name of α-effect. The value α is similar to such values of electrodynamics as dielectric and magnetic permeability and conductivity. It is difficult to know what Maxwell would say about the α-coefficient, because ε and μ are scalars, and α is a pseudoscalar. α is built as an average from the scalar product of speed and vortex: $\alpha = 1/3 \tau$ ($V \mathrm{rot} V$), where τ denotes turbulence memory time, and ($V \mathrm{rot} V$) is a mean value sign. Such characteristic is known in physics where it is called helicity.

Helicity is not found in traditional macrophysics, its appearance is due to the operation of averaging the turbulent velocity field. The mechanism, which causes the appearance of helicity in rotating stratified medium, is represented by the action of Coriolis force leading to additional winding of vortexes of a definite sign, and this creates helicity. Through the notion of helicity Parker describes two current loops (or two magnetic fields which take part in the dynamo process) in the following way. Let us represent an axisymmetric geomagnetic field as sum of poloidal magnetic field

B_P, for example field of a magnetic dipole, and toroidal magnetic field B_T, the magnetic lines of which are somewhere in the liquid core. Then the differential rotation ω makes a toroidal magnetic field out of the poloidal one, and helicity α creates a poloidal magnetic field out of the toroidal one. If α and ω are considerable enough and the inductive effect prevails over dissipation, then the magnetic field grows and dynamo works. Such mechanism is called $\alpha\omega$-dynamo mechanism. A toroidal magnetic field may be created out of the poloidal one by helicity as well. In this case, it is a question of α^2 dynamo (it means that the role of the differential rotation can be neglected) or $\alpha^2\omega$ dynamo (in this case both contributions must be considered).

There are models of nearly axisymmetric dynamo. Initially models of nearly axisymmetric dynamo developed in a simplified version, neglecting for simplicity the presence of the hard inner core and regarding the whole central zone of the Earth as occupied by convective flows. Presently the inner core managed to take place in this pattern, which led to considerable increase of the model's inflexibility and consequently approximated its predictions to the reality. Nowadays within models of nearly axisymmetric geodynamo it is possible to reproduce almost any element of the real evolution of the magnetic field. For example, Cox's scale of geomagnetic polarity reveals fractal properties. It means that the question of how many inversions of geomagnetic field have taken place within the last 160 million years is not formulated well enough. It points to the fact that different generalizations of the scale dramatically change the total number of inversions, and at actually achievable time intervals it is impossible to hope that paleomagnetic data really resolve the majority of the inversions. In this situation instead of trying to carefully calculate the number of inversions in the age of frequent change of polarities, it is more sensible to characterize the structure of the scale with the help of some auxiliary quantity which characterizes the frequency of the inversion recurrence and which is called fractal dimension.

The described difficulties of the models of nearly axisymmetric geodynamo led to the appearance in the 90-ies and rapid development of a fundamentally different approach proposed by P. Roberts and G. Glatzmaier. They abandoned the efforts of the theory of nearly axisymmetric dynamo to theoretically describe and parameterize as many elements of geodynamo as possible with the help of a comparatively small number of parameters.

Instead, they propose to relate as many details of geodynamo work as possible to the field of numerical modeling, and not to try to observe or evaluate the convection properties, but to receive them within the same numerical experiment as the growth of the geomagnetic field itself. Such attempts had been made before in different problems of fluid mechanics, but it was only then that such an approach brought absolute success.

Basing on the principal equations of geodynamics and electromagnetism and not going beyond the minimum of incoming information, Roberts and Glatzmaier received the configurations of the magnetic field, which surprisingly resembled the real configurations, they reproduced the evolution of the magnetic field as far as the first inversion and even before it. The calculation required using maximum abilities of the modern computers. The publication of the results did not only attract the attention of scientists, but also of general public.

The effect of the results on the scientific society was so fundamental, that they deeply discussed the question whether the problem of the formation of the geomagnetic field had theoretical content, after this series of works.

However, the very question of how the new results fit into the framework of the models of nearly axisymmetric dynamo requires theoretical study, and nowadays the plan of it can only be outlined. The calculations in question do not at all contain the level of description where the notion of helicity appears, and the correlation with the models of nearly axisymmetric geodynamo will first of all require restoration of this level.

The above mentioned information presents a contracted survey of the article "Geodynamo" and models of generation of the magnetic field"[1].

---

[1] Sokolov D. "Geomagnetism and aeronomy". V. 44 № 5 Moscow, 2004

Speaking about the energy source of geomagnetic dynamo, based on energy evaluations of kinematic models of dynamo, S. Braginskiy states that the basic cause of convection in the outer core is gravitational differentiation due to the differences in impurity concentrations, rather than heat release. According to this developing model of large-scale convection, the most active fractional separation of substance takes place in the border zone between the outer core and the mantle.[1]

The cause of repolarisation of the Earth dynamo according to Braginskiy may be in sudden outer impacts, for example if the origin of the convection is in the descend of heavier material from the mantle into the core then there can be sudden fluctuations in the ingress of such material. Such perturbations may be caused by inner instabilities and fluctuations in the core. The observed increase of repolarisations during the period of higher geological activity hints at the outer cause of the repolarisation.

The existing theory "Geodynamo" does not determine the origin of the 0,2° drift of the magnetic dipole per year to the west.

The application of the dynamo theory for the Sun, which is a gas star with nuclear reactions in the interior, to the Earth with its molten iron core, is hypothetical, as well as the energy source of geomagnetic dynamo and the causes of repolarisations of the Earth's dynamo[2].

The author proposes an alternative approach for explanation of physical processes in the Earth interior, on the basis of the laws of electrodynamics. He regards the appearance of the magnetic field of the Earth due to the transfer of the charge, which is in the core of the Earth, by the Earth rotation.

Geophysical research

The magnetic field of the Earth is of global importance for the life on the Earth.

---

[1] Ushakov S. "Physics of the Earth". V.1, Moscow: "Vinity", 1974.
[2] Braginskiy S. "Physics of the Earth" V.10. Moscow, 1972.

Geophysics explains that the appearance of the magnetic field of the Earth is due to the transformation of the kinetic energy of the movements of the molten iron in the Earth liquid core into magnetic energy.

Geophysics does not determine the causes of:

- the 0,2° per year drift of the magnetic dipole to the west;
- the drift of plates and continents.

The basic characteristics of the planet Earth shown in table 1 are established by geophysical researches.

Table 1

Parameters	Outer Core	Inner Core	Mantle
Depth interval, km	5150 – 6371	2900-5150	0-2900
Chemical compound, %	Fe-73;Ni-8; FeS-19	$Fe_2 0 – 100$	MgO -31; FeO – 13,1; SiO_2 - 47
Temperature, °C	5100	3800	0 ÷ 3800
Pressure, mln atm	3,3 ÷ 3,67	2,93 ÷ 3,33	0 ÷ 1,34
Electric conductivity $Ohm^{-1} Cm^{-1}$	3×10^5	$3 \times 10^3 - 3 \times 10^5$	$10^{-4} ÷ 10^2$

The rotation period of the Earth is $T = 24 \times 3\,600 = 8,64 \times 10^4$ s.

It is known that high temperature lowers the magnetizability of iron, which in such conditions can only be supported by external magnetic field. At temperatures $T > T_k$ S-valent and 3d-electrons of atoms of iron form mixed Fermi-liquid.

Biot-Savart law

The magnetic field of the Earth is created by the electric currents which appear at the transportation of the electric charge, located in the liquid core of the Earth, by the rotation of the Earth. If the electric charge consists of electrons and rotates together with the Earth, then the direction of the electric current will be opposite to the rotation of the Earth, i.e. clockwise. The magnetic induction inside the Earth core will be directed from the North pole to the South.

At calculating the magnetic field the physical Biot-Savart law for ring electric current was used: $dB = \dfrac{k_0}{c^2} I \dfrac{dl}{R^2}$ (see Appendix 1).

The magnetic field of the Earth appearing in the core creates an electric field, which moves electrons from the centre of the core to the mantle. Considering the difference of conductivities of the iron core and the silicate mantle electrons accumulate in the core layer from R = 2748 km to the mantle R = 3471 km, forming a layer of an electric charge. The average radius of the electric charge layer is R = 3148 km.

The calculation of the magnetic induction through the charge of the sphere of the average radius of the electric charge layer gives a result comparable to the calculation of the induction though the electric charge layer. Applying Biot-Savart law for sphere [1] we determine the magnetic induction in the centre of the Earth core: B = 2,58 Gs.

The calculated negative charge of the electric charge layer at the moment creating the magnetic field of the Earth, appears to be excess over the positive charge, which is in the centre of the core. The negative and the positive charges of the Earth core, as well as the magnetic induction, change during the cycle between the inversions of the magnetic field.

The given calculation of the magnetic induction in the core does not show the distribution of the magnetic induction throughout the core, it only determines the magnetic induction in the centre of the Earth.

The difficulty of making a precise calculation of the magnetic field of the Earth is determined by nonlinear processes, which take place in the Earth core, also by relative data about the chemical composition of the Earth core, the density, temperature and pressure in the Earth core, about the distribution of iron in the Earth core.

We know about the drift of the magnetic dipole of the Earth about 0,2° per year longitudinally westwards, which is measured by geophysicists of many countries. The axis of the magnetic dipole does not coincide with the axis of the Earth rotation, but is at the angle of 11,5° to the axis of the Earth rotation, and the centre of the dipole axis is 468 km away from the centre of the Earth. It indicates that the iron in the core is not symmetrically distributed in relation to the axis of the Earth rotation, and the configuration of the core is not defined. The calculation of the electric charge has been made for a round core with symmetrical distribution of iron in relation to the axis of the Earth rotation.

All the data taken from geophysical information for the calculation of the magnetic field of the Earth have a certain degree of precision, and the calculation based on these data can only be made with an equal precision.

For all further calculations the calculated value of the induction of the magnetic field of the Earth, existing nowadays, is used.

Induction currents in the core of the Earth

The magnetic field causes the appearance of electric currents. The rotation of the electric charge layer together with the Earth creates a non-rotating magnetic field. According to Faraday's law of electromagnetic induction, the rotation of the Earth core in the magnetic field causes the appearance of the electric field E_{mf}.

"If the electric field itself is conditioned by electromagnetic induction from alternating magnetic field, then its order of magnitude contains a superfluous multiplier $1/c$ [2] in comparison with the magnetic field". [Landau, 1992].

In our case there is a movement of the core in a constant magnetic field, which conditions presence of the electric field. If the magnetic field inside the core is directed from the North pole, the rotation of the Earth is counter-clockwise, then the electric field is directed to the axis of the Earth rotation.

As a result of the radial electric field activity, electrons will flow from the centre of the core to the electric charge layer.

Taking into account the high conductivity of the core it is possible to believe that the central part of the core presents a positively charged field. The rotation of the positively charged central part of the core also creates a magnetic field, which is inversely directed to the main magnetic field $E_{mf\ inv}$, as well as electric field $E_{ef\ inv}$. The main magnetic field will prevail over the inverse one until a certain time, according to the superposition principle.

Amplification of the magnetic field by iron

The magnetic field is amplified by iron, which is in the mantle and in the core. The magnetic permeability of iron at temperatures lower than Curie point (1043°K) $\mu \gg 1$, magnetic susceptibility $\chi > 1$. At temperatures higher than Curie point iron becomes paramagnetic, and the magnetic susceptibility corresponds to Curie-Weiss law: $\chi = c / (T - T_K)$, where c is a substance constant. The magnetic permeability $\mu = 1 + \chi$. Thus the iron of the core is an amplifier of the magnetic field.

The magnetic field is amplified by the iron in the mantle including minerals: magnetite and others, which are ferrimagnetics. At T lower than T_K for ferrimagnetics $\mu > 1$, at $T > T_K$ ferrimagnetics become paramagnetics and their magnetic susceptibility corresponds to Curie-Weiss law.

As the iron in the core is not distributed symmetrically in relation to the axis of the Earth rotation, it is obvious that the electric charge will be higher where there is more iron.

At lowering of the magnetic field and the approach of the inversion of the magnetic field, the magnetic poles start moving in the direction of the areas where there is more iron in the core of the Earth.

The inversion of the magnetic field

The rotation of the electric charge layer creates also an external magnetic field. Its lines of force incorporate with the internal field. The external field is weaker than the internal one, but it also creates the electric field directed away from the electric charge layer. As a result of the activity of this field electrons from the mantle will also flow to the electric charge layer.

When $E_{mf\ inv} = E_{mf}$; $B_{inv} = B$, there will be an inversion of the magnetic field of the Earth. The action of the electric field E_{ef} will begin. Electrons from the layer of the electric charge will flow to the central part of the core and to the mantle. It is obvious that the reduction of the negative charge will be greater than the reduction of the positive charge of the central part of the core. There will appear an inverse magnetic field B_{inv}, which will later be amplified by the iron of the core and mantle. There will be growth of the electric field $E_{mf\ inv}$, as a result of its action electrons will flow from the electric charge layer to the centre of the core.

This inversion will go on rather quickly, because the core has charge. Electrons from the electric charge layer will start flowing to the positively charged part of the core.

After the inversion of the magnetic field the flow of the electrons from the electric charge layer will continue till the next inversion of the magnetic field. The magnetic field of the positively charged core will decrease until the residual charge of the electric charge layer creates a magnetic field exceeding the inverse field.

Later on, there will appear an electric field E_{mf}, and the formation of the electric charge layer will begin, that will determine the basic magnetic field of the Earth.

MOVEMENT OF LITHOSPHERIC PLATES

The version of academician Korovyakov

In 1976 academician N. I. Korovyakov discovered by means of modeling the conditions and processes, going on in the centre of the Earth, a previously unknown regularity of eccentric drift of the inner core inside our planet. He writes: "The dense core does not at all stand regally in the middle of the globe, nailed there by authorities of geophysics, it travels in the melt of magma along a pentagonal trajectory". In his opinion, the movement of the core and the melt of magma along the perimeter of a pentagon influences the displacement of continents, growth of mountains, and drift of the magnetic poles of the Earth. The displacements cause earthquakes, tsunami, volcano eruptions, and influence the climate and the ocean currents[1].

Theory of tectonic catastrophes

The problem of tectonic catastrophes is officially recognized as global problem number one. The greenhouse effect and the catastrophes were discussed once again at "Earth summit", which took place in 2004 in SAR. The earthquake in the Indian Ocean, which took place December, 26, 2004, and the subsequent tsunami, which killed over 300 000 men, agitated the whole world. Do we know all about the life of the planet we call the Earth? Disasters on the Earth are often associated with the activity of the Sun, the interposition of planets and the Moon. Suppose they are not alone to influence the weather-formation and the catastrophes. In 1976 the UNO and UNESCO addressed the developed countries with a proposal to find out the causes, better say to specify the existing theory of the origins of the natural tectonic catastrophes. The physics of the focus are considered to be known, the mechanism of earthquakes is thoroughly studied, and the foreshocks are numerous. It was only a question of how to determine the place and the date. An option of creating a net of

---

[1] Simonov V. "Earthquakes, tsunami, catastrophes", Moscow, "Eksmo". 2011

stations in areas of seismic activity is regarded. The signal must be transmitted from them to the satellites, and from the satellites to superpower computers, which would calculate when and where it would "shake". Some studies however stumped the science. The physics of earthquakes were fundamentally wrong. Some unknown mechanism was in action. Some little-known process of disturbing all the geophysical fields and mediums was going on. "To put it plain, seismologist of the whole world pull their governments' leg and extort money from them for nothing. Hitherto there hasn't yet been a single correct forecast of an earthquake. The humanity is absolutely defenseless before this natural disaster" (academician I. N. Yanitskiy). The origins of earthquakes and volcano eruptions are one of the unknown mysteries of nature. Numerous studies do not offer a full and complete picture of the processes going on in the interior of the Earth. Such natural phenomena as gravitation, magnetism and spiral vortex are not clear to the modern science, and that is why it is so difficult to describe these processes. That's why we use several physics and several geometries. A huge mathematical apparatus which is used to describe natural processes, is often based on philosophy, which is downtrodden because of the existing opinion that "the truth is incognizable".

Lithospheric plates and their driving forces

As we know, the crust of the Earth – a part of the lithosphere – is a hard shell of the Earth which consists of the crust of the Earth and the upper part of the mantle. The total thickness of the lithosphere varies from 50 to 200 km. The lithosphere is divided into approximately 12 big plates, and several small ones drifting in the asthenosphere. Asthenosphere is a "molten" shell of the Earth, it is situated lower than the lithosphere in the depths range of 50 to 700 km. The inverted commas in the word "molten" are intentional. Nobody knows for sure what asthenosphere consists of. Nobody has got samples of asthenospheric matter. That's why there's not much probability in stating that it is molten rock, or magma. We can agree that under the hard crust there is a rock layer in the molten state, and we can probably measure its width, but the modern science does not know what is under it, besides the question about the existence of the "hard" core is not yet clear. That's why we will conventionally name the substance, which composes asthenosphere, magmatic. Though magma, which erupts from volcanoes, is not the substance of the asthenosphere. The deformation forces appearing in the Earth crust on the borders of the plates cause earthquakes at their independent moving. So runs the postulate, which serves as the basis of the theory describing the origin of earthquakes. However the theory does not fully explain the origin of the driving forces.

Let us ask a naïve, even childish question: who or what moves these plates, or can they independently move on their own? Probably many people have ever watched an ice drift on a river. Do the ice-floes independently bump, climb up each other, and turn over on their own? The majority will unambiguously answer, that it is the current of the river, that moves the ice-floes. Relating this answer to the interior of the Earth we conclude, that it is the power of the currents spreading under the hard crust in asthenosphere, that moves the tectonic plates. And the plates themselves FLOAT on asthenosphere. Then there is a question of how these currents are formed and how they move. Is it only due to convectional currents in asthenosphere? What is the nature of hydrodynamics of these currents? If we thoroughly scrutinize the form of the existing continents, we'll see that they resemble tails of stark vortex flows. The being formed areas drifted from the sub-polar areas towards the equator under the influence of vortex flows, Coriolis forces and gravitational forces. We can suppose with high probability, that within the life-time of the Earth there have been at least three such global drifts of lithospheric plates from the poles to the equator.

Let us pay attention to the Sun, which hasn't yet got the crust, and the energy impulses, which erupt from the interior, are easily registered like flashes, and the cooling areas are registered like black spots. And how do the processes going on inside the Earth differ from those in the Sun? Practically, the only difference is that we cannot observe them clearly because of the hard crust. But it doesn't mean that there are no similar flashes and spots. We just have to learn to see them. Similarly to the ejections that we observe on the Sun, the impulses' energy erupted from the Earth interior are huge, and they are registered by meteorological satellites at definite intervals of the electromagnetic emission of the Earth. Under the influence of the impulses of forces, which are formed in the interior, an immense fire-spitting mass of currents moves inside the rotating globe, first rising to the hard crust and cooling down, then going down to the "reactor" and moving on after getting another supply of energy.

The form of the inner surface of the lithosphere

What is the form of the inner surface of the tectonic plates? Can it be "smooth"? Even simple deduction gives a negative answer. Seismic studies also deny this "fact". The inner surface of the hard Earth crust cannot be smooth because of the presence of mountains. What role do mountains play? The answer is paradoxically simple. Mountains are refrigerators. Being considerably voluminous, mountains have greater thermal capacity, than the flat surface. Mountains grow down due to more intensive cooling of the existing and the incipient matter in the foot area, like hoarfrost grows in the freezer of the domestic refrigerator. That's why the Earth crust must have an uneven surface on the inner side of it as well.[1]

A specific character of Icelandic volcanoes

The volcanoes of Iceland belong to the so called fissure type. It means that the eruption does not come out from a single crater, but from a fissure, which in fact means a chain of craters. That's why their influence on the climate and the inhabitants of the Earth is much greater and more long-lasting, than that of the volcanoes of the central type – with one or several craters, even if they are very powerful, such as Etna, Vesuvius, Krakatau, etc., says leading researcher of the geography faculty of Moscow State University, candidate of geographical sciences Yury Golubchikov.

---

[1] http:// Теория тектонических катастроф. Шендеров В. И.

So, the eruption of the volcano on the Indonesian island Krakatau in 1883, which lasted 3 months, destroyed the bigger part of the island and caused deaths of 36 thousand people. And in 1783 the Icelandic volcano Lucky – though without causing so much destruction, was so harmful to the climate, that it led to much more victims. Within 7 months a great deal of fluorites (salts of hydrofluoric acid) and sulfurous gas was ejected from a 25-km-long fissure. Acid rains and a huge cloud of volcanic dust, which hang over the whole Eurasia and some regions of Africa and North America, caused such climatic changes, which led to crop failure, loss of cattle and starvation not only in Iceland, but also in other countries of Europe and even in Egypt. As a result of it the population of Iceland decreased by a quarter, and the population of Egypt got 6 times less. The crop failures and starvation years which followed the eruption, gave rise to social discontent. So, Lucky can be partly called "the initiator" of the French revolution of 1789 and the subsequent social disturbances, says Yury Golubchikov. In ancient times eruptions of the Icelandic volcanoes were still more tremendous. Scientists believe that they could have caused the extinction of mammoths and groups of animals connected with them, as well as the forest destruction in Iceland.

The volcano, which caused so much disturbance in Europe today, is 50 times smaller than Lucky, its fissure is "only" 500 m long. It doesn't even have a name, and is called after the glacier, under which it is situated, Eyjafjallajokull. However, even being so small, it spread real panic. Scientists remind, that the previous eruptions of this volcano always preceded the eruptions of another subglacial volcano, Katla, which is more active. If it happens again, the consequences may be disastrous. Another reason why the Icelandic volcanoes attract the attention of scientists is that the North Atlantic can be called "the cook-house" of our weather. Energy unloading of the interior of the Earth runs through Iceland, says docent of the chair of general and applied geophysics of Dubna University, candidate of science Vladimir Krivitskiy.

Will the "Icelandic clip" stand?

Member of Museum of physical geography of Moscow State University Ph.D. Nikolay Zharvin believes, that future of a considerable part of our planet depends on the condition of the Earth crust on the territory of Iceland and Greenland.

An unstable Greenlandic glacier once freezes, once melts at a period of several thousand years. The lithospheric plate under it sometimes sags, sometimes flattens. Since the beginning of the contemporary interglacial period the size of the Greenlandic glacier has considerably shortened, and it continues melting. Some European and American scientists believe, that within the XXI century the Greenlandic glacier will completely melt down. As a result, the Earth crust will rise approximately one kilometer up.

Besides near Greenland there is Iceland, which is crossed by a part of the global rift system – cracks in the Earth crust on the border on two lithospheric plates. The width of the rift valley under Iceland is about 100 km. according to the theory of the continent drift, the rift on the ocean floor is not static. Its width in the Atlantic Ocean grows by 2 cm a year. In the Pacific Ocean this movement goes on at the speed of 10-15 cm, and it makes 20 cm a year on the territory of Nazca plate.

There is not such expansion in Iceland, but the stability of such "clip" is undetermined. When the melting of the Greenlandic glacier and the flattening of the Earth crust reach the critical value, this clip can break. The North American plate will at that rush up, and the Eurasian one will tumble down. A huge crack will form, in which there will be a rush of water. When it contacts the matter of the upper layer of the mantle as hot as 1000-1200°C, there will be a great explosion, because water expands 1700 times at heating. As a result, the disposition of the lithospheric plate may take several hundred meters, or even a kilometer. Besides, a huge column of volcanic ashes will erupt into the air, so permanent night will set in. the heated water will rush up causing enormous tsunamis, which will not only destroy the coastal cities of North America and Europe, but also all life within hundreds and thousands of kilometers away from the coast.

The scientist considers it necessary to start designing a plan of evacuation of the dwellers of Murmansk, Kaliningrad, St.-Petersburg, and Archangelsk in case of such a disaster. Moreover, this tremendous explosion will not be limited to the territory of Greenland and Iceland, but will spread along the whole global rift system. There will be a global natural catastrophe, which will be followed by rise of unprecedented seismic activity. The majority of the volcanoes will wake. The evaporated oceanic water will rise up to 20-30 km and, spreading in the atmosphere of the whole Earth, will cause a biblical power rain.

Nikolay Zharvin thinks, that the "end of the world" may with great probability happen in 2030-2070. But he admits that this question has not yet been well studied. More realistic forecasts require making the analysis of the Greenlandic glaciers by drilling from coast to the centre and comparing the samples of the phytome. The site where a great temporal differential is found, could have been the place of the previous critical value.

In the scientist's opinion, in the coming summer the volcanic activity in Iceland may grow, because the Greenlandic glacier will continue melting and the deformation of the Earth crust will increase.

Mysteries of Iceland

Processes which take place on the territory of Iceland raise many scientific questions. So, leading researcher of the Museum of physical geography of Moscow State University, Ph.D., professor Mikhail Rudin points out, that two great anomalies of water surface, approximately 62-meter high, were discovered on the surface of the Atlantic Ocean between Iceland and Britain in the course of different tests. Scientists cannot yet explain this phenomenon.

Besides, near Iceland there is an oceanic current, which is not subject to the Coriolis force connected with the rotation of the Earth. To the south of this island the warm North-Atlantic current is abandoned by the Irminger Current, which doesn't set to the east, like the main current, but to the west, washing the coast of Iceland and

Greenland. Nikolay Zharvin supposes, that it might be due to the fact, that the entry of the oceanic water into the rift has begun.

Why does the seismic activity grow?

The growth of the seismic activity, which can be lately observed, is explained by scientists by several reasons.

In Nikolay Zharvin's opinion, any considerably great eruption of a volcano can lead to the chain reaction of eruptions and earthquakes all round the Earth. He compares an eruption to a rocket. When there is an explosion in the Earth interior, and an immense eruption rushes into the atmosphere, then an equally powerful blow will be directed downwards, and then sideward, causing a chain of cataclysms.

Vladimir Krivitskiy thinks, that the tectonic activity of the Earth is connected with its rotation round the Sun, and the rotation of the Solar system in the galaxy. He explains, that we exist in a current of elementary particles, which constantly affect the inner state of the Earth. The force of this impact can vary. So, when the Earth moves towards the Sun, the current of elementary particles, which are constantly generated in the Sun, produces a much greater effect on it (the so called Doppler effect). As a result, the tectonic activity of the Earth increases. Probably that's why it is in spring, that the activity of many volcanoes increases greatly. Besides, the geoenergetic activity of the Earth increases under the influence of the solar activity. The eruptions and earthquakes of late years may be connected with flashes on the Sun and ejections of solar matter, which have become more frequent. The scientist notices, that the volcanic activity is determined by the inner energy of the Earth.

Mikhail Rukin explains the growth of the seismic activity by the reconstruction of the inner build of the planet, and links it with the alternation of the Earth orbit around the Sun towards the expansion of the conventional ellipse. The scientist says, that the gravitational influence of the Sun causes a specific pulsation of the Earth. When the planet approaches the Sun, it as if "breathes out", and when it moves away, an "inbreathing" takes place. It explains the monthly and the annual regularity of the increase of the tectonic activity of the Earth. Mikhail Rukin believes, that the present growth of the seismic activity is a regular natural process, not the beginning of a global catastrophe. The reconstruction of the inner build of the Earth is to finish in 2012.

Real volcanoes instead of the mystical warming

Scientists believe, that exploration and monitoring of volcanoes is much more important, than the mystical problem of the global warming. The human influence on the climate is probably grossly overestimated. Meanwhile, the tectonic processes may be a real menace. That's why it is necessary to conduct regular tests of the seismically dangerous zones, using both seismic and neutron sensors. In Russia potentially dangerous zones are Caucasus with its dormant volcano Elbrus, Baikal, where the process of the formation of a new fracture in the Earth crust is taking place, and Kamchatka, where volcanoes are the highest mountains in the world. The height of Kamchatka volcanoes measured not from the sea-level, but from the bottom of the Kuril-Kamchatka Trench, is about 12 thousand meters, which is much higher than the Himalayas. At the same time the Kamchatka volcanoes are competitive with Icelandic ones, taking into account the power of their influence on the climate.[1]

The increase of tectonic activity

One of the most tremendous earthquakes took place in 1952 within the Kuril-Kamchatka insular arc. It was followed by tsunami, which practically washed away the town of Severo-Kurilsk.

In 1960 a 9,2-magnitude earthquake took place in Chile, the rupture length being ~ 800 km, and with the area of 100 thousand km^2.

March, 11, 2011 – a 9-magnitude earthquake and the following tsunami happened near the coast of Honshu Island in Japan. Its epicenter was located

---
[1] http:// Специфика исландских вулканов

approximately 370 km away from Tokyo at the depth of 25 km. First came vigorous underground shocks, and then a gigantic 10 up to 20 meter high wave rushed down at the speed of 750 km/h on the coast of Japan. The wave swallowed the coastal towns of the north and north-east of Japan. Thousands of people died. Thousands of houses were destroyed; 6 million people were deprived of electricity and water. The city of Kessenuma, with the population of 70 thousand people, was enveloped in fire. Petroleum refineries burnt down. Breakages occurred on two nuclear power plants: Fukushima-1 and Fukushima-2.

All those earthquakes happened in the basin of the Pacific Ocean. The Pacific plate is surrounded by several lithospheric plates, which bump into the Pacific plate and glide beside it. At the collision of the oceanic and continental plates the oceanic plate sinks under the continental one and disappears in the mantle. In these places deep-water trenches are formed.

What is the real origin of the force, moves lithospheric plates which so powerfully and causes earthquakes?

Brief information on the existing theory of the plate tectonics

Nowadays the movement of lithospheric plates is explained with the help of the theory of the plate tectonics.

The theory of plate tectonics is based on 3 fundamental theses:

1. The mechanical model of the upper mantle consists of an elastic hard outer layer – lithosphere, and an underlaying mouldering layer – asthenosphere. The oceanic lithosphere is defined as an external cold border layer of convective cells of the upper mantle.

2. For the division of the lithosphere into plates summaries of the seismicity of the Earth are used.

3. Lithospheric plates are hard bodies, that's why the kinematics of their movement on the sphere submits stringent geometrical rules.

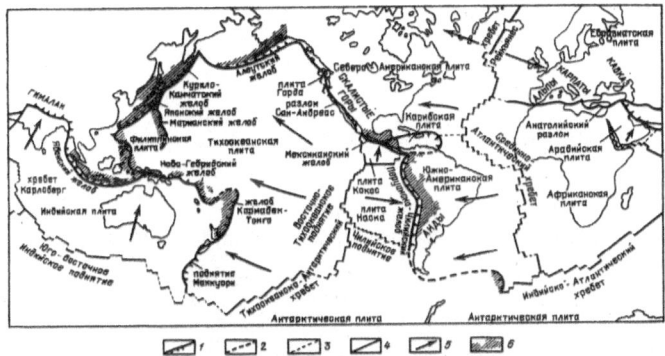

The Mercator map of the Earth surface. The lithosphere breaks into large hard plates, each of them moving as a whole (acc. to J. Dewey). The arrows show the movement of the plates assuming that the African plate is static. The borders of the plates are marked by earthquake belts. The plates drift away from the axes of the mid-oceanic ridges, glide by each other along the transform fractures and bump into each other in subduction zones. 1 – subduction zones, 2 – borders of the plates drawn indefinitely, 3 - transform fractures, 4 – axes of the ridges, 5 – directions of the plate movement, 6 – areas of deep-focus earthquakes.

Picture 1.

The Mercator map of the Earth surface depicted on Pic.1 [Zharkov, 1983] shows the division of the Earth lithosphere into plates with continents located on them. Every plate moves as a whole. The arrows show the direction of the plate movement.

Plate tectonics is a surficial demonstration of the convective movement inside the mantle. A fundamental peculiarity of the Earth tectonics is linearity of its main structures – mid-oceanic and continental rift systems and deep-water trench systems. It results from the structure of the developed convection in the upper mantle, which being as it is, conditions the formation of linear ascending currents – the layers which outcrop in rift zones and create the outer thermal border layer – oceanic lithosphere. In the sites of oceanic trenches lithospheric plates begin sinking into the mantle. At the plates sinking volcanism and earthquakes take place.

The speed at which lithospheric plates move has been measured, and the direction of their movement has been determined. The measurement of the plates speed has made it possible to formulate the following rules:

1) the more relative part of the plate's area is occupied by a continent, the lower is its speed;

2) the greater the relative length of the borders of plates absorption, the faster is their speed;

3) plates located in polar zones move slowly, and plates located in the equatorial area move faster.

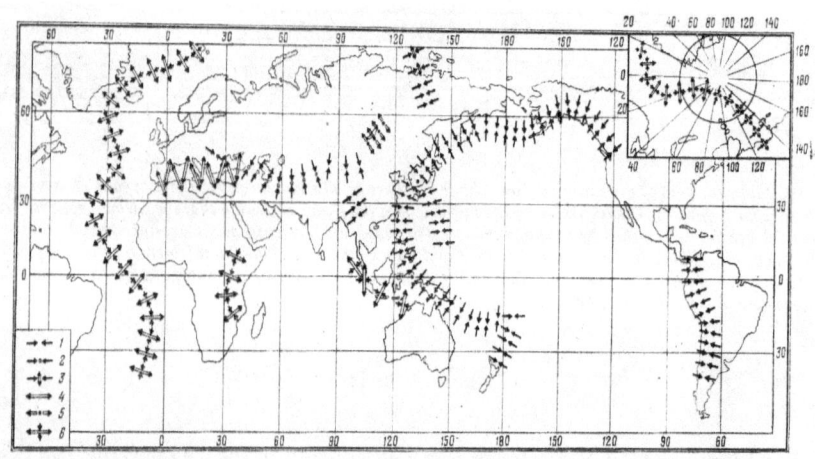

Orientation of the main strain axes in the elastic strain fields of the Earth. The peak compression strains (1, 2,3) or tensile strains (4, 5, 6) are oriented horizontally or transversely to the strike of the structures.

Picture 2

Picture 2 shows orientation of the main strain axes of mid-oceanic and continental rift systems.

The initial idea of the mechanism of plate tectonics, which leading specialists in plate tectonics stuck to during first years after its creation, stated that hard lithospheric plates are entrained by mantle currents. Nowadays leading specialists tend to believe, that the movement of the lithosphere, which is part of a large-scale convection in the upper mantle, entrains at plate sinking the underlaying asthenosphere due to the force of viscous friction. Thus, it is not the asthenospheric flow, that moves lithospheric plates, but vice versa, it is lithospheric plates, that drive the viscous asthenosphere and undergo its braking force.[1]

Plate tectonics does not offer an explanation of the main direction of the lithospheric plates moving westwards.

The author assumes it possible to regard the mechanism of plates movement as both, the idea of the mantle currents unsupported by the driving force, and a large-scale convection at plates sinking, unsupported by the driving force as well; and he proposes an alternative approach to explaining the driving force of the mechanism of the plates movement on the basis of physical laws of electrodynamics.

Drift of the Earth core

We know about the drift of the magnetic dipole of the Earth 0,2° per year longitudally westwards. Radial currents interact with the basic magnetic field, which results in the rotating of the core. It agrees with the laws of electromagnetism.

According to Ampère law, at the interaction of the radial electric currents with the magnetic field of the Earth a great force is created: if the magnetic field inside the core is directed to the drawing, and the electric currents are directed to the axis of the Earth rotation, then the force action is directed clockwise (westwards). Movement of the earth core makes one turn round the axis of the Earth in the mantle in the west direction for 2000 years.

The calculation of the driving force of the core drift mechanism is shown in Appendix 2.

At resistivity of a column, which is 1 meter long and has the section of $1м^2$, $\rho = 3,3 \times 10^{-6}$ Oнм, we calculate, that the force, which is formed at the radial currents flowing from the axis of the core rotation up to $R = 3471$ km, and which conditions the drift of the core, makes $3,5 \times 10^{12}$ N.

---

[1] Zharkov 'The inner structure of the Earth and planets", Moscow, Nauka, 1983.

Drift of the magnetic dipole of the Earth and the lithospheric plates

The drift of the magnetic dipole of the Earth 0,2° per year longitudally westwards is measured by geophysicists of many countries. The magnetic dipole makes a turn around the axis of the Earth for about 2000 years. The rotation of the Earth core determines the westward drift of the magnetic dipole of the Earth.

The axis of the magnetic dipole is displaced in relation to the axis of the Earth rotation. It is obvious, that the elemental composition of the core and the geometrical dimensions of the core are determined conventionally, and the uneven spread of iron, which is highly conductive and has magnetic properties, determines the displacement of the axis of the magnetic dipole in relation to the axis of the Earth rotation.

The interaction of the radial electric currents with the magnetic field of the Earth is the cause of formation of a powerful force of $3,5 \times 10^{12}$ N, which turned the Earth core westwards, and the driving force of the mechanism the movement of lithospheric plates.

During the inversion and pre-inversion condition of the magnetic field, when the magnetic induction and the radial electric currents are minimal or vanishing, the force, which determines the drift of the earth core, reduces to zero and the Earth core stops drifting. It removes the driving force of the lithospheric plates movement for the period of the inversion of the magnetic field of the Earth. After the inversion the magnetic field changes the poles, the induction of the inverse magnetic field will start growing, there will be an increase of the radial electric currents, directed from the centre of the core, and the core will again start drifting in the direction inversion to the rotation of the Earth. It is obvious, that due to the stop of the drift of the Earth core during the inversion of the magnetic field, the driving force of the lithospheric plates movement changes its condition, and it can result in the growth of geological activity and the increase of the amount of earthquakesю

The substance of the Earth bark, if it somewhere forms a mountain range, plunge deeply into heavy mantle masses. Figuratively speaking, mountains have roots sinking inward. Mountains are like icebergs – their major part is in the depth. These peculiarities are supported by detailed seismic studies.

Along the western coast of South America there stretches a mountain ridge – the Andes, and along the western coast of North America there is a mountain ridge of the Cordilleras.

The westward drift of the core involves the lower layers of the mantle and the protuberances of the bases of the lithospheric mountain ridges.

In the rift valley of the Mid-Atlantic Ridge lava erupts and cools down, forming the oceanic lithosphere and moving in the parties from the Mid-Atlantic Ridge.

The plates with continents North America and South America drifted several thousand kilometers away from Europe and Africa during 200 million years, and there was formed the Atlantic Ocean.

The Atlantic Ocean expands at the expense of the Pacific Ocean. The North-American and South-American plates pressure upon the Pacific plate, and as a result the plates Cocos and Nasca under them and the Pacific plate itself plunges into the mantle at a collision with the lithospheric plates located on the west of the Pacific plate. These zones of the Pacific basin are characterized by intensive volcanic activity and earthquakes.

Deceleration of the Earth rotation

The core of the Earth drifts in the mantle in the direction counter to the Earth rotation. The angular velocity of the rotation of the Earth core is 0,2° less than the speed of the Earth rotation, i.e. the Earth core does not make a turn of 2π = 6,2831852 radian a day, but 6,2831757 radian a day. 0,2° a year = 0,0005749° a day = 0,0000095 radian a day.

The alternation of the kinetic energy of the Earth core at the rotation of the core counter the rotation of the Earth, is shown in Appendix 3.

The alternation of the kinetic energy of the core makes $7,456 \times 10^{22}$ J.

The total kinetic energy of the Earth equals to the sums of the energies of component parts, every second the Earth energy would lose $7,456 \times 10^{22}$ J, even if the Earth were a completely hard body. But the fused mantle, which has high dynamic viscosity (coefficient of internal friction), absorbs the greater part of the energy of the core rotation. The difference of the angular velocities of the Earth and the core of the Earth is: $\Delta\varpi = \omega_s - \omega_я = 7,272205 \times 10^{-5} - 7,272194 \times 10^{-5} = 1,1 \times 10^{-10}$ radian/second.

The linear speed of the core rotation relative to the mantle on the border between the core and the mantle is $V = \Delta\varpi \times 3,471 \times 10^{6} = 3,818 \times 10^{-4}$ m/sec = 33 m/day = 12 km/year.

At its drift the core carry away the mantle and the lithospheric plates. The researches held by New Zealand and American scientists in deep-water parts of the Pacific Ocean with the help of hydrophones, registered gurgling, hissing, crackle and long "groans". These are sounds produced at asthenosphere slipping under the lithosphere.

In present days geophysicists explain the slowdown of the Earth rotation to the tidal forces of the Moon. The friction of the sea tidal waves against the floor in shallow seas is determinative, and leads to a great angle of lag of earthly tides $\delta = 2$–$4°$. The tidal friction leads to deceleration of the Earth rotation and the Moon's moving away from the Earth. The Earth loses its moment of momentum and the kinetic energy of rotation. The friction of the sea tidal waves against the floor in shallow seas is not comparable to the constant friction of the asthenosphere against the bottom of the whole lithosphere, when the asthenosphere, which has high viscosity, catches the protuberances of the mountain grounds and different cavities in the foot of the lithosphere, and moves the lithospheric plates.

It is the friction of the asthenosphere against the lithosphere, that causes the principal deceleration of the Earth rotation. In 1991 duration of the astronomical day increased by 1 second, in July 1992 the amendment added one more second to the day, and in 1993 already 2 more seconds were added.

THE THERMAL ENERGY OF THE CORE

The global warming

Global warming is a process of gradual increase of the average annual temperature of the atmosphere of the Earth and the World Ocean in the XX and XXI centuries. The viewpoint of the Intergovernmental Panel on climate change (IPCC) of the UNO, coordinated with National Academies of G8 countries, is that the average temperature on the Earth has increased by 0,7°C since the beginning of the industrial revolution (the second half of the XVIII century), and that "the greater part of the warming, which has been observed in the recent 50 years, is due to the human activity", first of all to the emissions of gases, which cause the greenhouse effect, such as carbon dioxide (CO_2), and methane (CH_4). Evaluations received by the climatic models, which the IPCC refers to, show, that in the XXI century the average temperature of the Earth surface can increase by the value of 1,1 up to 6,4 °C.[1]

According to the calculations of the research officers of Potsdam institute of climatic change studies (Germany) the level of the World Ocean will increase by 0,75 – 1,9 m to the year of 2100. Such an increase is extremely dangerous for Venice, Los Angeles, Amsterdam, Hamburg, St. Petersburg, San Francisco, and Lower Manhattan. Scientists' forecasts are as follows: every century due to water expansion at heating, the level of the world ocean grows by 1 m. The melting of the West Arctic ice shield will cause a 6 m rise of the world sea level, the melting of the Greenland ice shield will result in a 7 m rise, and that of the Antarctic ice shield will give a 61 m rise. An approximate chronology of submersion of territories can be made. In 100 years a 1 m rise of the sea level will cause Venice submersion. In 150 years a 2 m rise will submerge Los Angeles, Amsterdam, Hamburg, and St. Petersburg. In 200 years a 3 m rise will submerge San Francisco, and Lower Manhattan. In 350 years a 5 m rise will submerge South London. In 400 years a 6 m rise will submerge Shanghai and Edinburgh, and so on. Densely populated and developed coastal regions and

---

[1] http:// Глобальное потепление. Материалы из Википедии

lowland archipelagos, such as the Marshall Islands in the Pacific Ocean, the Maldives in the Indian Ocean, and some Caribbean states will suffer from submersion first of all. Except the rise of the world sea level, the increase of the global temperature will cause changes in the quantity and the distribution of atmospheric precipitation. It can result in the growth of natural disasters, such as floods, droughts, hurricanes, and so on, in the drop of crops harvest on the damaged territories and the growth of harvest on the other territories (due to the increase of the density of carbon dioxide). The warming may probably intensify the frequency and distribution area of such phenomena.

Some scientists believe, that the global warming is a myth, some of scientists deny the possibility of man's influence on this process. There are scientists, who do not deny the fact of the global warming and admit its anthropological origin, but do not agree, that the most climatically ruinous factor is the industrial emission of greenhouse gases.[1]

The greenhouse effect

The greenhouse effect was discovered by Joseph Fourier in 1824, and was first quantitatively studied by Svante Arrhenius in 1896. This is a process, at which the absorption and emission of infrared radiation by atmospheric gases causes the heating of the atmosphere and the surface of the planet.

The main greenhouse gases on the Earth are: water vapour (accounts for approximately 36-70 % of the greenhouse effect, clouds not included), carbon dioxide (CO_2) (9-26 %), methane (CH_4) (4-9 %), and ozone (3-7 %). The atmospheric densities of CO_2 and CH_4 have increased by 31 % and 149 % correspondently, as compared to the beginning of the industrial revolution in the middle of the XVIII century. According to some researches, such levels of density have been achieved for the first time during the late 650 thousand years – a period, for which reliable facts from samples of polar ice have been obtained.

---

[1] Serov M. "Global Warming". Moscow, «Knizhniy Klub Knigovek», 2010

About half of the greenhouse gases, which appear in the course of the economical activity of the humanity, remain in the atmosphere. About three quarters of all the anthropogenic emissions of the greenhouse gases during the late 20 years have come as a result of the extraction and burning of oil, natural gas and coal. The greater part of the rest of the emissions is due to the change of the landscape, first of all deforestation. The theory of the anthropogenic contribution to the modern climate change as a result of the greenhouse gas emission can be supported by the facts, that the observed warming leads first of all to the growth of average temperatures in high (sub-polar) latitudes, to the growth of average temperatures in winter in middle latitudes, and to the decrease of the night cooling. It is also true, that rapid heating of troposphere layers happens against the background of less rapid cooling of stratosphere layers. On the other hand, the effect of the greenhouse gases is grossly overestimated. The volume of the atmosphere of the Earth is about 15milliard km^3, which is 18375000 billiard tons; at the emission of about 9 milliard tons the density of the gases makes 0,00005%. At such densities any global climate changes are hardly possible.

The human activity

Results of the late researches confirm the theory that the cause of the global warming is in the human activity. The research which was carried out by scientists from Scotland, Canada and Australia showed, that the probability of natural, non-anthropogenic origins of climate change on the planet does not make more than 5 %. According to this research, since 1980 the average air temperature has risen by 0,5 degrees centigrade, and the Earth keeps heating at the speed of 0,16 degrees a decade.

Change of the solar activity

Various theories have been proposed to explain the temperature changes of the Earth through the corresponding changes of the solar activity. The IPCC's third report claims, that the solar and volcanic activities could account for half of the temperature changes before 1950, but after this their total effect was practically zero. In particular, according to IPCC, since 1750 the influence of the greenhouse effect is 8 times higher, than the influence of the change of the solar activity. Later studies evaluated the influence of the solar activity on the warming after the year of 1950. Nevertheless the conclusions remained approximately the same: "The best estimate of the warming from solar forcing is estimated to be 16% or 36% of greenhouse warming...". ("Do Models Underestimate the Solar Contribution to Recent Climate Change?" Peter A. Stott, Gareth S. Jones, and John J. B. Mitchell «Journal of Climate», Dec, 15, 2003). However there are some scientific works, that assume the existence of mechanisms strengthening the effect of the solar activity, what is not taken into account in the modern models; they also suppose that the importance of the solar activity as compared to other factors is underestimated. These theses are disputed, but they make a dynamic trend in research. The conclusions which will appear as a result of this discussion may turn out to play a key part in the issue: in what degree the mankind is responsible for climate change and what is the degree of natural factors in that issue.

Other versions of the warming on the Earth

There are many other explanations of the possible current increase of the average temperature of the Earth surface, disregarding the role of industrial greenhouse gases.

The observed warming lies within the limits of the natural climate variability and does not require a special explanation.

The warming is a result of the end of the cold Minor glacial period.

The warming has been observed for a too short period, that's why it is difficult to say for sure that it really takes place.

It must be taken into account, that except the influence of anthropogenic factors, the climate on our planet undoubtedly depends on many processes going on within the system "the Earth – the Sun – the Space". Except the collisions with large asteroids and comets, occasional, but multiple in the history of the Earth and catastrophic in their consequences, the Earth atmosphere experiences also regular impacts of planetary and cosmic origin. Such cycles can be divided into four groups:

"Superlong", each lasting 150–300 million years, are characterized by the most significant climate changes on the Earth. They are most likely to be connected with the period of the Sun's revolution around the mass centre of our Galaxy and the transits of the Solar system through the regions of the Milky Way with different density of gas-and-dust matter, which, depending on its composition, can both screen the Sun's emission, and intensify thermonuclear reactions in it.

"Long" cycles, connected with the tectonics of the lithospheric plates and the intensity of volcanic activity. They are firmly established in the paleontology chronicles, but are irregular in duration and last from several to tens of millions of years.

"Short" periods, the so called "Milankovitch cycles", lasting 93 000, 41 000 and 25 750 years, which occur due to periodic oscillation of perihelion of the Earth orbit and the orientation of the axis of the Earth rotation, which is determined by phenomena of nutation and precession. Of these two astronomic phenomena the periodical change of the angle of inclination of the Earth rotation axis to its orbit plane, i.e. nutation, comes first to influence the general insolation of the surface.

The last category is conventionally called "ultrashort" periods. They are connected with solar activity rhythms, among which there are presumable periods with duration of 6000, 2300, 210 and 87 years, except the undoubtedly existing 22- and 11-year-long cycles of the Sun activity.

Superposition of different in origin and duration periods of intensity change of solar radiation, which reaches our planet, together with thermal inertia of oceans, continents movement, volcanic activity and may be the influence of back reactions of the earth biosphere as a whole, determine the average temperature of the Earth surface and the distribution of climatic zones in different geological epochs. This multicomponent complex of a variety of alternating-sign geophysical and cosmic factors of influence upon the Earth climate, can condition – according to some scientists – the warming which is observed nowadays. The man is not yet able to influence processes of such a scale.

The global warming does not at all mean warming everywhere and at any time. In particular, in some place the average summer temperature may increase, and the average winter temperature may fall, that is the climate will become more continental. The global warming can only be revealed by averaging the temperature in all geographical locations and in all seasons.

According to one of the theses, the global warming will lead to a standstill or significant weakening of the Gulf Stream. It will cause a considerable drop of the average temperature in Europe (while temperature in other regions will grow, but not necessarily in all of them), because the Gulf Stream heats the continent by carrying warm waters from the tropics.

According to the hypothesis of climatologists M. Ewing and W. Donne in crioage there is an oscillation process, at which glaciation (ice age) originates from climate warming, and deglaciation (end of the ice age) originates from cold snap. It is connected with the fact, that at Cainozoe, which is crioage, at the melting of ice polar caps the amount of precipitations grows at high latitudes, which in winter leads to local increase of albedo. Later there is a drop of temperature in deep areas of continents of the North hemisphere with the following formation of glaciers. At the freezing of ice polar caps glaciers in deep areas of continents of the North hemisphere begin to melt, not receiving enough feed in the form of precipitations.

The negative changes in Europe are as follows: growth of temperatures and intensification of droughts in the south (resulting in the decrease of water resources and reduction of hydroelectric power production and agricultural production, worsening of touristic conditions), reduction of the snow cover and mountain glacier contraction, growth of flood and catastrophic river overflow hazards; intensification of summer precipitations in Central and Eastern Europe, increase of the frequency of forest fires and peat-bog fires, reduction of forest productivity; growth of instability of soils in North Europe. In the Arctic it is a catastrophic reduction of the area of the blanketing glaciations, the area of sea ice, intensification of the coast erosion.

Some researchers (for example, P. Schwarz, D. Randall) offer a pessimistic forecast, according to which in already the first quarter of the XXI century an abrupt climate jump in an unpredictable direction is possible, and it can lead to the beginning of a new ice age with the duration period of hundreds of years. The wide consensus of climatology scientists about the continuation of the growth of global temperatures led to some states, corporations and individuals trying to prevent the global warming or to adapt themselves to it. Many ecological organizations struggle for taking measures against the climate change, mainly by consumers, but also on the municipal, regional and governmental levels. Some of them call for the limitation of the world's production of fossil fuels, referring to the direct connection between fuel burning and CO_2 emission.

The Kyoto protocol

Today the main world convention about counteraction to the global warming is the Kyoto protocol (approved in 1997, ratified in 2005), an addition to The United Nations Framework Convention on Climate Change. The protocol includes more than 160 countries of the world and covers about 55% of the world's emissions of the greenhouse gases. The first stage of the realization of the protocol is to finish at the end of 2012. The international negotiations about a new convention began in 2007 on the island of Bali (Indonesia), and were continued at the UNO conference in Copenhagen in December 2009.

In 1980 over 100 million tons of CO_2 were emitted into the atmosphere in the eastern part of North America, Europe, the western part of the USSR and big cities of Japan. CO_2 emissions of the developed countries in 1985 made 74% of the total volume, the part of the developing countries was 24%. Scientists suppose that by the year of 2025 the participation of the developing countries in carbon dioxide production will rise up to 44%. Russia and the countries of the former USSR have lately reduced considerably the emissions of CO_2 and other greenhouse gases into the atmosphere. It is first of all connected with the changes going on in these countries, and the drop of the production level. According to the statement of president Dmitry Medvedev, Russia is going to support the EU initiative and will assume obligations to reduce CO_2 emissions by 20-25%.

In December 1997 at the meeting in Kyoto (Japan) devoted to the global climate change delegates from more than one hundred and sixty countries passed the convention obligating the developed countries to reduce CO_2 emissions. The Kyoto protocol obligates thirty eight industrially developed countries by 2008 – 2012 to reduce CO_2 emissions by 5% of the 1990 level. The European Union is to reduce the emissions of CO_2 and other greenhouse gases by 8%, the USA by 7%, Japan by 6%.

The protocol provides for a system of quotas of greenhouse gases emissions. Its point is that each of the countries (so far it refers only to the 38 countries, which assumed obligations to reduce emissions) gets the permission to emit a certain amount of greenhouse gases. At the same time it is assumed, that some countries will exceed the emission quotas. In such cases these countries or companies will be able to buy the right to extra emissions from those countries or companies, whose emissions were lower than the allocated quota. Thus it is assumed, that the main target – a 5% reduction of the greenhouse gases emissions in the next 15 years – will be achieved. There is also an international conflict. Such developing countries as India and China, which make a significant contribution to the atmosphere pollution with greenhouse gases, were present at the Kyoto meeting, but did not sign the agreement.

Criticism of the theory of the global warming

A prominent British botanist and broadcaster David Bellamy supposes that the main ecological problem of the planet is the reduction of the rainforest area in South America. In his opinion, the threat of the global warming is grossly overestimated, while the vanishing of the forests where there live two thirds of all the animal species and plants of the world, is a real and serious menace for the humanity.

A Russian theoretical physicist V.G. Gorshkov came to the same conclusion basing on a theory of biotic regulation, which had been designed by him since 1979. According to this, irreversible climate changes are most likely not to be caused by greenhouse gases, but by breakage of homeostatic mechanism of the global moisture and heat transfer, which is provided for by the vegetation of the planet, on condition of some overlimit reduction of the natural forest area.

A famous American physicist Freeman Dyson claims, that the measures suggested for fighting the global warming have long been far from scientific, but belong to intrigues and speculative business.

Founder of <u>Weather Channel</u>, journalist John Coleman believes that "the so-called global warming is the greatest scam in history". According to his words "Some dastardly scientists with environmental and political motives manipulated long term scientific data to create an illusion of rapid global warming. ... there is no run away climate change. The impact of humans on climate is not catastrophic. Our planet is not peril. ... In time, a decade or two, the outrageous scam will be obvious."

There were minor changes of the average temperature of the Earth during late 500 million years. During almost the whole history of the Earth the temperature was considerably higher than it is nowadays.

A Danish economist Bjorn Lomborg considers, that global warming does not have such a menacing character as it is depicted by some experts and journalists echoing them. «The issue of the warming is overheated», – he says. Lomborg's views are presented in details in the book «Cool it: The skeptical Environmentalist's Guide to Global Warming».

Professor A.P. Kapitsa, the member correspondent of the Russian Academy of Sciences, head of the chair of the Geography faculty of the Moscow State University, considers the contribution of the mankind to the climatic changes to be insignificant against the background of cosmic and geophysical factors.

A number of critics indicate that in the past (for example, in Eocene) the temperature was considerably higher than today, and though many species then died out, later the life prospered.

Climategate

In November 2009 as a result of actions of a group of unknown computer trespassers the mail server of the University of East England was cracked, and general public learned about the correspondence of scientists. Critics claimed, that from correspondence it may be concluded, that lately the temperature on the Earth has not been rising. The authorities of the University of East England propagated a declaration, in which the choice of the letters for the publication was called biased and "denying the world's admitted fact of a negative influence of the human activity on the climate". Also in response to criticism the Climatic Center of the University of East England in cooperation with Met Office Hadley Centre December, 7, 2009 distributed in free access part of the database, on the basis of which the temperatures of the Earth surface had been calculated (HadCRUT). The data represent a selection, which contains information from about 1500 ground weather stations out of the total number of 5000 stations. One of the most vivid processes connected with global warming is the melting of glaciers. For the late half a century the temperature in the southwest of the Antarctic, on the Antarctic peninsula has increased by 2,5 °C. In 2002 an iceberg of the area of 2500 km² broke off from Larsen ice shelf (that has the area of 3250 km² and is over 200 meters thick), located on the Antarctic peninsula. The whole process of destruction took only 35 days. Before that, the glacier had remained stable during 10 thousand years, since the end of the latest glacial age. Thawing of the ice shelf led to a release of plenty of icebergs (over a thousand) in the Weddell sea. Nevertheless, the area of the Antarctic glaciation grows. Acceleration of the process of the permafrost degradation has been registered. Since the early 1970-ies the temperature of permafrost soils in Western Siberia has risen by 1,0 °C, in Central Yakutia – by 1-1,5 °C. In the north of Alaska since the mid 1980-ies the temperature of the top layer of the frozen rocks has increased by 3°C.

Other aspects of climate change

The global climate change is not limited to warming. There are also processes of oceanic salt density change, air humidity growth, change of rainfall properties, and melting of the Arctic ice at the speed of 600 thousand square km per decade. The atmosphere becomes more humid, more rainfalls occur at high and low latitudes, and less in tropics and subtropical regions.[1]

North seaway will be free from ice

Nowadays the navigation period through the narrow place – the strait of Vilnitskiy – which separates the Taimyr peninsula from the Northern Land (Severnaya Zemlya) archipelago, and connects the Kara sea with the Laptev sea, takes on average 20-30 days. As a result of the global warming the strait might be navigable for 120 days by the end of the XXI century. Cargoes from Europe to Japan and China will be able to run through the North seaway.

At present the glaciers of Northern and New Land are rapidly collapsing. In future the glaciers of Alaska, and especially the ice shield of Greenland will also start collapsing more quickly. The destruction of ices on this island is especially obvious: from 1979 till 2005 their area reduced by 20%. Water leaks along the cracks down the glacier, and immense parts of it slide down the water lubricant into the ocean.

---

[1] http:// Глобальное потепление. Материалы из Википедии.

From the global warming to the global cold snap

One of the theories of the global warming leads to a paradoxical conclusion: the rise of temperature on the Earth will result in the global cold snap. Melting of the glaciers of Greenland will carry a huge mass of water into the Atlantic Ocean. In the Atlantic there is a current of "Gyre of Broker" which catches a huge amount of water flowing from south to north at the depth of several hundred meters. At the equator this water heats. The Gulf stream current is part of this gyre. When this water reaches the latitude of Iceland cold severe winds cool down the upper layers of the water, which causes rise of deep waters. The salty heavy cold water goes down and sinks nearly to the floor. And when it reaches the floor, it starts moving back to the south. This place in the north of the Atlantic where the water mass cools and goes down, is the mechanism, which rotates the immense oceanic circulation. If large amount of fresh water from the melting Greenland leaks to the North Atlantic, it will dilute the mass of the "heavy" water, which will stop being "heavy" and will not be able to sink to the floor. The conveyor which carries a great deal of warmth, will stop; North Europe and Scandinavia will be deprived of the warmth, which heats them now, a severe cold snap will come. [1]

A global cold snap is coming

A considerable cold snap followed by glaciation took place comparatively not long ago from the geological point of view, 25 thousand years ago. Thousands of years ago it changed into the global warming. An interesting thesis concerning this process was proposed by A.A. Barenbaum. According to his calculations it turns out, that some thousands of years ago, the Solar system approached Sirius. There were two suns in the sky. Sirius, located outside the Pluto orbit, was shining weaker, than the principal luminary, but much brighter than the Moon, and was transmitting rather much energy.

---
[1] Serov M. "Global Warming". Moscow, «Knizhniy Klub Knigovek», 2010

Some earth processes are influenced by movement of other planets of the Solar system. Not long ago ecologists and astronomers paid attention to the fact, that the movement of large planets – the Jupiter, the Saturn, the Uranus, and the Neptune – changes the mass centre of the solar system. If the Earth turns out to be between the Sun and large planets, the Earth climate gets warmer due to the currents of the solar energy, extracted from the star's interior by giant planets. Mass distribution in the Solar system can influence the position and configuration of the Earth core.

Climate is to a great extent determined by the Sun and the interaction in the Earth-to-Sun system.

Since the end of the XX century the trend of the warming changed for the cold snap.

But the ices of the Arctic go on melting rather fast. A global catastrophe caused by this is not to be expected, and the world sea level will not rise. Unlike the Antarctic, the ices of the Arctic are floating. According to Archimedean principle, floating ice releases as much water, as it contains. So, melting of the Arctic ices is a benefit for our economy and transport.

From 1979 to the beginning of the new millennium the area of the ice shield in the Antarctic grew from 14,6 million to 15,9 million square kilometers. As the northern glaciers float, and the southern ones rest on the continent, at further continuation of these processes the water level in the ocean can drop, but not grow as apologists of the global warming claim.

What awaits the Earth in the foreseeable future?

One of the best-reasoned models was proposed by specialists of the Chief astronomical observatory of Russian Academy of Sciences. They suppose that the low temperature period will shift into the warming only at the beginning of the XXII century.

Based on the cycling of the minor glaciations, it is quite possible. The amount of the energy emitted by the Sun has been decreasing since 1990-ies, and will reach the minimum in approximately 2041. Climate is connected with the Sun with some lateness. That's why even taking into account the humanity's influence and the continuing industrialization, the temperature will be falling for some time. The thermal energy of the world ocean will somewhat slow down this process, but the cold will last some more decades. The human activity doesn't influence the climate much. The main influence comes from the Sun-and-Earth connections, which determine quite a number of cycles of climate change.

In any case, no global catastrophes are to be expected.[1]

Explosions are to blame for the global warming

Among the causes of the global warming such phenomena as explosions are mentioned (if we do not speak about volcanic explosions during eruptions). The author of the "explosion" theory is a Russian researcher V. Shenderov, who has been studing climatic phenomena for 30 years. According to his theory explosions, which occur during hostilities, construction and mining works, produce a great effect on the interior of the planet. According to Newton's laws, enormous energy of multiple explosions absorbed by the Earth crust is sure to arouse the counteraction, which is represented by the Earth climate changes. V. Shenderov claims, that 0,04 % CO_2 of the atmosphere is not able to cause such a large-scale melting of glaciers, as we observe nowadays. The real cause of the disasters lately becoming more frequent are explosions of different purposes.

It is explosions that cause multiplication of the number of hurricanes, melting of the permafrost, slipping down of glaciers of Greenland and the Antarctic (as a result of the water sheet formation under the glaciers). The basic proof of the theory is a predominantly floor-level character of melting of glaciers and the permafrost. [2]

---

[1] Sapunov V.. "A global cold snap is coming" St. Petersburg. "Astrel-Spb". – 2010.
[2] Simonov V. "Earthquakes, tsunami, catastrophes", Moscow, "Eksmo". 2011

The planetary climatic hypothesis

The Earth civilization was formed in the age of the so called holocenic interglacial period. It began approximately 10 thousand years ago, and will finish – according to mathematical models – at the end of the III thousand A.D., that is in approximately 3000. Since that moment another global cold snap will begin, which will reach the climax after 8000. The basic argument of the planetary climatic hypothesis is the fact of a periodical shift of the tectonic activity in rift valleys. Convectional currents in the interior of the Earth agitate the Earth crust with different force, and that leads to the existence of such ages. Geologists possess materials, which convincingly prove that climatic fluctuations are chronologically linked to the periods of the maximal activity of the interior. In 2002 American scientists and the NASA determined on the basis of laser probing from the American space satellites, that during the recent years the radius and the form of the Earth have been changed, in particular, the Earth has become more flattened from the poles and more widened on the equator. This led to redistribution of voltages inside the Earth, causing the rise of the seismic activity we observe nowadays. Rift valleys are numerous on the ocean floor, where the crust is very thin and breaks easily under the pressure of convectional currents. In these areas the volcanic activity is very high. Here the mantle matter is constantly erupting from the interior.

According to the planetary climatic hypothesis, it is the magma eruptions, that play the key role in the fluctuation process of the historical reformation of the weather regime[1].

On the floor of the Pacific Ocean scientists of California Institute discovered a new type of volcanic activity. Lots of mini-volcanoes (activity spots) erupt fire-spitting lava from cracks of the Earth crust. The existing statistic data show, that earthquakes are becoming more and more frequent. In the XX century (1900-1930) 2000 earthquakes were registered. From 1940 to 1982 there happened 1000 earthquakes annually! In 1983 300 000 earthquake shocks were registered, which

---

[1] Serov M. "Global Warming". Moscow, «Knizhniy Klub Knigovek», 2010

means 800 per day. Since 1984 the number of the registered earthquakes is 1000 per day.[1]

Milankovitch cycles

In the XX century a Serbian astronomer M. Milankovitch developed a daring theory explaining the basic climatic events by laws of the Earth rotation, and fluctuations of the axial incline towards the orbit. Now the planet's axis has the angle of 23,5° towards the plane of the ecliptic. In summer (from June) the Northern hemisphere gets more light than the Southern one. In the northern latitudes days become longer, and the position of the Sun is higher. Meanwhile it is winter in the Southern hemisphere, days are shorter, and the Sun is lower. Half a year later the Earth travels along its orbit to the opposite side of the Sun. The axial incline remains the same. Now the summer is in the Southern hemisphere, days are longer and there is more light, and in the Northern hemisphere it is winter. In the Northern hemisphere there are more continents, more land. On the land during the cold snap huge glaciers are formed, which increase the amount of the reflected sunlight (the planet's albedo). The illuminance of the northern latitudes considerably changes due to variations of the astronomical parameters. In the real solar system the Earth is not alone to revolve around the Sun. it is influenced by the gravitation of the Moon and other planets, and this gravitation produces a little, but important effect on both the orbit and the rotation of the Earth. This effect is represented by the following:

Precession. The axis of the Earth does not always retain the same position; it moves slowly in a circular cone. The axis of this cone is perpendicular to the plane of the Earth orbit, and the angle between the axis and the generatrix of the cone is 23,27°. This effect is called "precesion". The action of the gyroscope is based on it. When the gyroscope starts moving, it rotates quickly around its axis, while the axis itself circumscribes a cone. The same happens to the axis of the Earth, and the period of a complete revolution is approximately 26 thousand years. Now the Earth is so

---

[1] Simonov V. "Earthquakes, tsunami, catastrophes", Moscow, "Eksmo". 2011

inclined, that in January (when the Earth is in the closest approach to the Sun) the northern hemisphere is turned away from the Sun. In 13 thousand years the situation will change for the opposite: in January the northern hemisphere will be turned towards the Sun, and January will become midsummer in the northern hemisphere.

Nutation. In addition to the slow precession of the Earth, the angle of inclination of the Earth axis slightly fluctuates as well. These fluctuations are called "nutation". Now the axis has the inclination of 23° to the plane of the Earth orbit. Every other 41 thousand years under the influence of both the Moon and Jupiter (a far-away but huge planet) the angle of inclination reduces to 22° and then grows again till 23°.

Change of the orbit shape. Because of gravitation of other planets the shape of the earth orbit changes in the course of time. From an ellipse prolonged in one direction it changes into a circle, and then into an ellipse prolonged in a different direction, perpendicular to the original one; then another circle, and so on. This cycle lasts about 93 thousand years. Milankovitch believed, that the Earth climate is influenced by all these three cycles, each of them being connected with a certain astronomic effect. When they intensify each other, a cold snap and a glaciation period can be expected. But normally these three factors act in different directions and their impacts do not merge together, so the climate returns quickly to its normal state. Some doubts about Milankovitch hypothesis are raised by the fact, that the climatic changes caused by the change of the Earth orbit usually take several tens or even hundreds of thousands of years. The fast enough climate change we observe now, is probably due to the influence of some other facts besides those indicated by Milankovitch.

The Great glaciation

The Earth is at present in the gripe of glaciation. "At present" means a period of relative warming, an interglacial interval, but for several previous millions of years our planet was on average colder, than during the major part of its history. Nowadays there are ice caps near the south and north poles, the size of more than a continent.

Origins of the periodical cold snaps, which the Earth underwent, are evidently rather complicated and not quite clear, in spite of decades of explorations of this phenomenon. But details of the very late Ice Age in which we live now, become more and more documented. Such phenomena as volume fluctuations of ice mass, sea level measurements, the reaction of land vegetation to climatic changes, and even actual temperature fluctuations within several latest millions of years – all these facts are already well known. The man, who is most of others associated with general assuming of the idea of a continental-scale glaciation is Louis Agassiz, a Swedish scientist who collected information about ice deposits round the whole Europe, and later in North America. Under the influence of the facts Agassiz understood, that a considerable part of North Europe used to be buried under a thick layer of snow. Works of Agassiz and other scientists showed, that North Europe, the bigger part of Great Britain, Canada and the northern half of the USA were buried under a several-kilometer thick ice layer in not far-away geological past. Oceans retain the most unbroken chronicles of climate changes at glacial period.

Even in the tropics, far away from direct influence of the polar ice caps, sediments reveal features which are closely connected with cycles of coming and going of glaciers. In fact, it was not before long cores of sea sediments got available for exploration, that it became possible to interpret the real details of the Great Ice age. Though these sediments carry much indication of ice climate change, probably a more valuable indicant is the chronicle of oxygen isotopic composition of sea water. Organisms living in the ocean and building shells from calcium carbonate store in the isotopic composition of oxygen the characteristics of the sea water around them, by which they register the signal reflecting both the temperature of the water and the amount of the water which was connected with glaciers' ices.

Graph 1 shows regular changes of isotopic composition of oxygen in shells of bottom-living organisms, which reflect changes of the ocean temperature and ice volume within the late 600 thousand years.

Graph 1

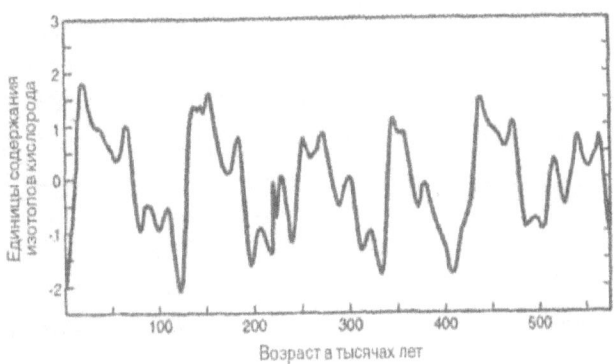

Positives values in this graph correspond to cold, glacial periods, and the negative ones - to interglacial periods. In this graph there are several interesting peculiarities. One of them is regularity: for the late half a million years the content of oxygen isotopes in water recurs in an amazingly systematic way, reflecting the existence of cycles in coming and going of glaciers. Here only five ice ages are depicted, but if we continue this graph back in the past to about 3 million years, the graph's character turns out to remain the same! It indicates the existence of periodical interchange of cold and warm periods. The duration of the cycles depicted in the graph, makes roughly 100 thousand years. For more ancient parts of the graph these cycles are somewhat shorter, but in spite of this, it is obvious that some factor is very regular in affecting the Earth climate. There is a certain rhythm in the sequence of ice ages, which must be governed by the influence of some unknown factor, which changes in a similar way. [1]

---

[1] J. MacDougall "Short History of Planet Earth". Transl. from English, St. Petersburg. "Amfora", 2001

Brief information on the existing theory of emission of the Earth's inner heat

According to seismic data the capacity of the Earth crust is on average 30-40 km. The Earth crust on continents can be averagely presented as a 15-kilometer layer of granite resting on an equally thick layer of basalt, which are sources of radiogenic heat emission. Concentration of radiogenic sources of heat in granites is $1,9 \times 10^{-5}$ cal/cm^3 a year, in basalts $0,35 \times 10^{-5}$ cal/cm^3 a year. The contribution of the both layers in the heat flow from the Earth makes 34 cal/(cm^2 a year). If we compare the derived figure with the average heat flow which lifts from the surface of the Earth and makes ~ 47 cal/cm^2 a year, then we will see that it is by 70% determined by heat emissions in the granite and basalt layers. Later on, this value was reduced to 40 %.

The Earth crust on the oceans consists of a 5-6- kilometer basalt layer, and the value of the heat flow coincides with the value of the flow on the continents. The explanation of this result is based on the supposition, that the number of radiogenic sources of heat per unit of area is equal on both the continents and the ocean. The difference is only in the fact, that on the continents the sources are mainly gathered in the outer granite and basalt layers, and on the oceans these sources are dispersed to the depth of several hundreds of kilometers.

The heat flow from the Earth is estimated $Q = 2,4 \times 10^{20}$ cal/year. The energy of radioactive disintegration is estimated in the following way. The heat emission in the Earth per unit of mass is $Q_M \sim Q/M \sim 2,4 \times 10^{20} / 6 \times 10^{27}$ (the mass of the Earth) = 4×10^{-8} cal/(g x year), which practically coincides with the heat emission per gram in coaly chondrites, from the material of which our planet is considered to be formed. Heat loss by the Earth for the period of its existence of ~ $4,6 \times 10^9$ years at its presumably constant heat flow is $Q_t = 2,4 \times 10^{20}$ cal/year x $10^9 \approx 1,1 \times 10^{30}$ cal. Now let's estimate the heat capacity of the Earth. The heat capacity of silicates is ~ 0,3 cal/(g x °C), the heat capacity of the "iron" matter in the core is three times lower : ~ 0,1 cal/(g x °C). Consequently, the masses of the silicate cover and the Earth core make 4×10^{27} g and 2×10^{27} g. Then the average heat capacity of the Earth $c_з \sim 0,3 \times 4 \times 10^{27} + 0,1 \times 2 \times 10^{27} = 1,4 \times 10^{27}$ cal/°C. Dividing the average heat loss by its average heat capacity $c_з$ we will determine the "effective" temperature of the Earth cooling: $\sim Q_т / c_з = 1,1 \times 10^{30} / 1,4 \times 10^{27} \sim 800°C$, which can be called little or moderate, but hardly much, as compared to its initial heating.

Geophysics explain the heating of the core by the initial heat which appeared at the formation of the Earth ~ 1000 °C, and the energy of gravitational differentiation of the primary homogeneous Earth into the iron core and the silicate mantle and crust~2500 °C.[1]

The existing model of the cooling Earth does not take into account the inner heat of the Earth, emitted by the Earth core at flow of the radial currents; that's why the question of the cooling of the Earth is suppositional.

The thermal energy excreted by the Earth core at flow of the radial currents

Powerful thermal energy excreted at flow of the radial currents in the Earth core determines the heating of the core and the Earth on the whole.

According to Joule-Lenz law, the amount of the warmth excreted for time t at electric current flowing $Q = I^2Rt = IUt = Pt$.

The calculation of the thermal energy emitted by the core is given in Appendix 4.

At resistivity of a column with the length of 1m and section of $1m^2$: $\rho = 3,3 \times 10^{-6}$ Ohm, we calculate: the thermal energy excreted by the Earth core at radial electric currents flowing from the centre of the core to R = 3471 km, makes $8,2 \times 10^{10}$ watt/sec = $1,97 \times 10^{10}$ cal/sec = $6,92 \times 10^{17}$ cal/year.

The mantle of the Earth is a potent heat insulator.

The excreted thermal energy of the core produces a great influence on the Earth climate.

---

[1] Zharkov 'The inner structure of the Earth and planets", Moscow, Nauka, 1983.

The successions of the glacial periods, in the author's opinion, are determined by periods of inversions of the Earth magnetic field.

Powerful thermal energy excreted at flowing of radial electric currents in the core of the Earth does not only fuse the core, but also heats the mantle of the Earth. At inversions and preinversional conditions of the magnetic field, when the radial electric currents are weak or vanishing, excretion of heat by the liquid core decreases, starts cooling of the Earth and a cold glacial period begins. The accumulated thermal energy of the Earth and the inertness of the Earth cooling should be taken into consideration.

In the rift valley of the Mid-Atlantic Ridge, which has numerous volcanoes, lava is constantly erupted, thus heating the water of the Atlantic Ocean. The warm Gulf Stream, which heats the north of Europe and Canada, is well known to everybody. The ocean bottom heats waters of the oceans as well. The heat of the oceans is the cause of formation of huge mass of water vapor leading to increasing of the precipitation level and hurricane formation.

Melting of permafrost, slipping of glaciers in Greenland and Antarctic confirm the fact, that the warmth of the Earth's crust determines melting of glaciers and permafrost.

Postamble

The Earth core is the least studied part of the Earth due to its size and the remoteness from the Earth surface.

The exactness of the geophysical data is mainly conjectural. The physical processes which go on in the Earth core are also poorly investigated.

The Earth rotation, which transports the electric charge in the core, determines the appearance of the magnetic field. The radial electric currents appearing in the Earth core, in the course of rotation of the iron core in the magnetic field, determine the heating of the core and the Earth on the whole, and increase the charge in the electric charge layer, which later leads to the inversion of the magnetic field.

Interaction of the radial electric currents with the magnetic field of the Earth is the cause of formation of a powerful force, determining the drift of the Earth core westwards, i.e. inversion the Earth rotation, and the driving force of the mechanism of the lithospheric plates movement. The drift of the core explains the drift of the magnetic dipole 0,2 degrees a year longitudally westwards, and determines the slowdown of the Earth rotation.

The Earth core is situated inside the Earth, between the parallel of 33° northern latitude and the parallel of 33° southern latitude. The maximum heating of the surface of the Earth and the oceans takes place in this area. Polar zones (those of the South and North poles) get less heat. This prosess influences the climate of the polar zones and the preservation of ice caps on the poles of the Earth as well.

It must be noted, that rising of the fused core temperature leads to temperature increase of the mantle and the lithosphere, which results in melting of the lower layers of the lithosphere as well. The consequences of that are so far unpredictable.

The drift of the core westwards involves the lower layers of the mantle and the founding of the lithosphere at low latitudes as much as possible. The movement of the lithospheric plates at high latitudes is insignificant, which is also indicated in the theory of "Tectonics of plates": the plates located in polar zones move slowly, and the plates located in the equatorial zone move faster.

Nowadays we deal with the process of global warming on the Earth. The Earth has accumulated enormous thermal energy, and the accumulation is still going on. It is obvious, that the rise of the temperature in the liquid core is the reason of the mantle temperature rising, as well as increasing of the thermal stream from inside the Earth. The geothermal gradient at the Earth surface is estimated at 20 grad/km. The greenhouse effect increases the temperature on the Earth surface thanks to heat retention by different components of the atmosphere, such as carbon dioxide and methane.

The general attention to the problem of the global warming on the Earth deserves respect.

Due to melting of continental glaciers the sea level has risen by ~ 120 m for the late 20 thousand years. Melting of the glaciers of Greenland and the Antarctic can rise the sea level by ~ 50 m.

Monitoring of changes of geothermal gradient at the Earth surface, and monitoring of the change of the greenhouse gases quantity in the atmosphere allows to determine the extent of the influence of these two factors on the global warming.

An organized monitoring of the magnetic field of the Earth can probably provide a timely enough warning, so that to carry out a strategy of protection from of the inversion of the Earth magnetic field.

At present the induction of the magnetic field of the Earth is reducing. There is information, that the southern pole of the magnetic field is drifting towards the Indian Ocean, and the northern magnetic pole is moving towards the East Siberian magnetic anomaly through the Arctic Ocean. At reduction of the magnetic field and uneven distribution of iron in the core, this is logicality and signifies the coming inversion of the magnetic field of the Earth.

All the calculations conducted on the basis of physical laws, are approximate, and only support the truth of the theory of "The Magnetic Field of the Earth". The calculations do not take into account: that all the processes in the Earth core are non-linear; geophysical data have a conjectural character; the size and configuration of the core, and the distribution of iron in the core have not yet been precisely defined; the exact chemical composition of the core and its conductivity are neither known.

The drift of the magnetic dipole in of inversion the direction the rotation of the Earth is a confirmation of the drift the core of the Earth the inversion the rotation of the Earth, and of the correctness of the theory of "The Magnetic Field of the Earth".

The author of the book «PHYSICS» Jay Orear (Cornell University), translation from English Moscow "World", 1981, page 283, an example 9, at calculation uses formula $U = VB$ (without multiplier $1/c$). At similar calculation of the driving force of the mechanism of the Earth core drift $F = 10{,}5 \times 10^{20}$ N, the thermal energy emitted by the Earth core $P = 62{,}28 \times 10^{33}$ cal/year.

Epilogue

1. There is the global warming on the Earth, though the amount of the thermal energy from the Sun has not increased.

In summer 2012 due to the lack of rains there was a drought in the basin of the Amazon River. The drought involves 30 % of the area of evergreen forests, which are "the lungs of the planet". Climatologists explained it by rise of the temperature of the Atlantic Ocean waters.

The character of melting of the tundra from below and the Arctic Ocean ices confirms the rise of the temperature of the Earth surface and the world ocean due to the heat coming out of the Earth. Melting of the ice of Greenland, the Antarctic and other glacial systems can significantly raise the world sea level, which will lead to flooding of many territories, the population of which will have to be moved.

2. The magnetic field of the Earth creates around the Earth the magnetosphere, which protects life on the Earth from solar and space radiation. At present the induction of the magnetic field of the Earth reduces. The inversion of the magnetic field of the Earth is coming. During the inversion the magnetosphere will vanish, and it will be necessary to protect people and animals from the radiation. There may be variants of individual protection, and collective protection, when an artificial magnetic field is created over a certain area.

3. During the inversion of the magnetic field of the Earth, when there is no magnetic induction, and no radial electric currents the driving force of the lithospheric plates will disappear. The Earth core will stop drifting westwards. The North American and South American plates will stop moving westwards. The crack in the Mid-Atlantic Ridge shut, and the excretion of lava, that heats waters of the Atlantic Ocean, will finish. The warmth of the Gulf Stream heating the north of Europe and North America will disappear It is beginning Ice Age. A new migration of peoples will begin.

Slowdown of the Gulf Stream did take place in the past. As climatologists discovered, the warming, which began 12 700 years ago was suddenly – in tens of years – replaced by a cold snap in the Northern hemisphere, which lasted 1300 years. The cold snap ended as abruptly as it had begun. Supposedly then happen inversion of the magnetic field of the Earth. The Atlantis disappeared at that period, and there happened earthquakes and tsunami, which were several hundreds of meters the height.

4. Processes which go on in the Earth core are beyond the human power. The magnetic field of the Earth is created in the Earth core by the electric current of 1,9 milliard amperes, the thermal energy of the Earth core of $6,92 \times 10^{17}$ cal/year, which heats the core and the mantle, and the potent force of $3,5 \times 10^{12}$ N, which rotates the Earth core inversion the rotation of the Earth, confirmation this.

5. These questions require a thorough study and taking timely measures in order to avoid consequences of the inversion of the magnetic field of the Earth.

APPENDICES

Calculation of the magnetic field of the Earth (Appendix I)

The magnetic field of the Earth is created by electric currents which appear at the nsfer of the electric charge located in the core of the Earth by the rotation of the Earth.

At calculating the magnetic field the physical Biot-Savart law for ring electric current was used: $dB = \frac{k_0}{c^2} I \frac{dl}{R^2}$. $\frac{k_0}{c^2} = \frac{\mu}{4\pi} = \beta$; $\mu = 1,257 \times 10^{-6}$ вс/ам – is the magnetic constant, dl – is the vectorial length of the current element.

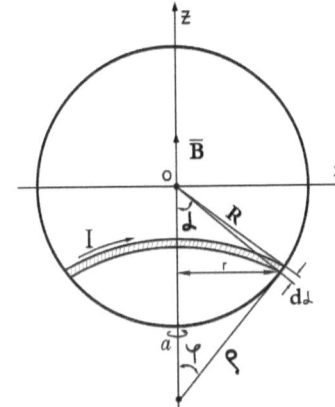

Pic .1

Picture 1 illustrates calculation of the magnetic field of the Earth and other calculations.

B – induction, I – ring electric current, R – radius of the sphere; r – radius of the ring; ρ – polar radius; a – distance between the pole and the centre of the Earth

Magnetic field component along axis Z: $dB_z = \beta I \frac{dl}{R^2} \sin\alpha$; $\sin\alpha = \frac{r}{R}$; then $dB_z = \frac{\beta I r}{R^3} dl$; We calculate integral by the ring $B_z = \frac{2\pi\beta I r^2}{R^3}$; B_z – induction of the ring electric current.

We calculate the induction of the electric current of a sphere:

$dq = \sigma ds$; $\sigma = \frac{Q}{4\pi R^2}$; dq – element of charge; σ – surface charge density; ds – element of area; Q – transferred charge; $ds = 2\pi r dl$; $dl = Rd\alpha$; $r = R\sin\alpha$; $\lim d\alpha \to 0$; α angle- the angle between radius R and axis Z.

$dq = \frac{Q}{4\pi R^2} 2\pi R \sin\alpha R d\alpha = \frac{Q}{2} \sin\alpha d\alpha$; $dI = \frac{dq}{T} = \frac{Q}{2T} \sin\alpha d\alpha$;

$$dB = \frac{2\pi\beta dIr^2}{R^3} = \frac{2\pi\beta Q \sin\alpha R^2 \sin^2\alpha}{2TR^3}d\alpha = \frac{\pi\beta Q \sin^3\alpha}{TR}d\alpha;$$

$d\alpha$ – determines the width of the electric current ring. The magnetic induction of the electric current of a sphere – B:

$$B = \frac{\pi\beta Q}{TR}\int_0^\pi \sin^3\alpha d\alpha = \frac{4\pi\beta Q}{3TR} = \frac{\mu Q}{3TR}; [1]$$

The magnetic field of the Earth, which appears in the core, creates an electric field, which moves electrons from the centre of the core to the mantle. Taking into account the difference of conductivities of the core and the mantle, electrons are accumulated in the layer from R = 2748 km to R = 3471 km. Average radius of a layer of electric charge R = 3148 km.

Applying Biot-Savart law and solving it in relation to the charge of the sphere with radius R = 3148 km, we determine the induction the magnetic field of the Earth.

$a = 6371$ km; ρ – polar radius; φ – angle between ρ and axis Z

$$dB_z = \frac{\beta \times I \times \sin\varphi}{\rho^2} \times dl \ ; \ \sin\varphi = \frac{r}{\rho}; \text{ then } dB_z = \frac{\beta Ir}{\rho^3}dl \ ; \ r - \text{ring radius}.$$

Integrating by the ring, we calculate: $B_z = \dfrac{2\pi\beta Ir^2}{\rho^3}$ – the induction of the ring electric current.

We calculate the induction of the electric current of a sphere:

$$dq = \sigma\, ds; \ \sigma = \frac{Q}{4\pi R^2}; \ ds = 2\pi\, r\, dl; \ dl = \rho\, d\varphi; \ r = \rho \sin\varphi;$$

$$dq = \frac{Q\rho^2}{2R^2}\sin\varphi d\varphi;$$

$$dI = \frac{dq}{T} = \frac{Q\rho^2}{2TR^2}\sin\varphi d\varphi; \ dB = \frac{2\pi\beta r^2}{\rho^3}dI = \frac{\mu Q\rho}{4TR^2}\sin^3\varphi d\varphi; \ \sin\varphi = \frac{R\sin\alpha}{\rho};$$

$$\rho^2 = a^2 + R^2 - 2aR\cos\alpha; \quad \rho\sin^3\varphi d\varphi = \frac{R^3\sin^3\alpha}{\rho^2} = \frac{R^3\sin^3\alpha}{a^2 + R^2 - 2aR\cos\alpha} =$$

$$\frac{0{,}778 \times 10^6 \sin^3\alpha}{1{,}259 - \cos\alpha};$$

$$dB = \frac{\mu Q x 0{,}778 x 10^6}{4TR^2} \times \frac{\sin^3 \alpha}{1{,}259 - \cos \alpha} d\alpha = 2{,}858 \times 10^{-19} Q \times \frac{\sin^3 \alpha}{1{,}259 - \cos \alpha} d\alpha;$$

$$B = 2{,}858 \times 10^{-19} Q \int_0^\pi \frac{\sin^3 \alpha}{1{,}259 - \cos \alpha} d\alpha = 3{,}575 \times 10^{-19} Q;$$

$$\int_0^\pi \frac{\sin^3 \alpha}{1{,}259 - \cos \alpha} = 1{,}251$$

Induction of the magnetic field of the Earth on the pole: $B = 0{,}6$ Gs, then $Q = \frac{B_{pol}}{3{,}575 \times 10^{-19}}$.

Calculation of the induction through the charge of a sphere with an average radius of the electric charge layer gives a result comparable with the calculation of the induction through the electric charge layer. Applying Biot-Savart law for sphere [1], we determine the magnetic induction in the centre of the Earth core: $B = 2{,}58$ Gs.

Calculation of the driving force of the mechanism of the Earth core drift

(Appendix 2)

According to Ampère law, at the interaction of the radial electric currents with the magnetic field of the Earth a great force is created: if the magnetic field inside the core is directed to the drawing, and the electric currents are directed to the axis of the Earth rotation, then the force action is directed clockwise (westwards). Movement of the earth core makes one turn round the axis of the Earth in the mantle in the west direction for 2000 years.

Let's determine the force, which conditions the drift of the core. According to Ampère law:

$F = IBl\sin \varphi$, where F stands for force (N), I – electric current (A), B – induction of the magnetic field (T), l – length of conductor (m), $l = r$, r – radius.

If angle φ between the induction vector and the direction of the current is 90 degrees, then F = IBl;

U = VBl, where U is tension (V), V – linear speed (m/sec), B – magnetic induction (T), l – length of conductor (m), l = r.

$V = \dfrac{2\pi r}{T}$, where T = 8,64 x 10^4s the period of the Earth rotation. $U = \dfrac{2\pi r^2 B}{T}$,

$I = \dfrac{U}{R} = \dfrac{2\pi r^2 B}{TR}$, where R is resistance of conductor (Ohm). We'll make a substitution: $F = \dfrac{2\pi r^3 B^2}{TR}$. $R = \dfrac{\rho_{y\delta} r}{ds}$, where $\rho_{y\delta}$ is resistivity considering the temperature ($\dfrac{kgm^3}{s^3 A^2}$), ds – conductor section, (m^2). Then $F = \dfrac{2\pi r^2 B^2 ds}{T\rho_{y\delta}}$. Function F (r) is distributed all over the volume of the core (ball). Since F = f(r), then $dF = \dfrac{4\pi r B^2 ds}{T\rho_{y\delta}}$

Let's calculate force F by the volume of the ball. In conformity [2] we will introduce coefficient 1/c for U.

$$F = \dfrac{1}{c} \iiint \dfrac{4\pi r B^2 ds}{T\rho_{y\delta}} \, dx\, dy\, dz;$$

We express the integral through spherical coordinates by the volume of the ball:

$$F = \iiint \dfrac{4\pi B^2}{cT\rho_{y\delta}} \rho \sin\theta\, \rho^2 \sin\theta\, d\theta\, d\varphi\, d\rho = \int_0^{2\pi} d\varphi \int_0^R \rho^3 d\rho \int_0^\pi \dfrac{4\pi B^2}{cT\rho_{y\delta}} \sin^2\theta\, d\theta.$$

Considered is, that multiplier $\rho^2 \sin\theta d\theta d\varphi$ is equivalent to sectional area of element ds in spherical coordinates.

$$F = \dfrac{4\pi B^2}{cT\rho_{y\delta}} \int_0^{2\pi} d\varphi \int_0^R \rho^3 d\rho \int_0^\pi \sin^2\theta d\theta = \dfrac{4\pi B^2}{cT\rho_{y\delta}} 6{,}28 \dfrac{R^4}{4} 1{,}571 = \dfrac{31 B^2 R^4}{cT\rho_{y\delta}} = 3{,}5 \times 10^{12} \text{ N}.$$

At resistivity of a column, which is 1 meter long and has the section of 1м², ρ = 3,3 x 10^{-6} ohm, we calculate, that the force, which is formed at the radial currents flowing from the axis of the core rotation up to R = 3471 km, and which conditions the drift of the core, makes 3,5 x 10^{12} N.

Calculation of the change of the kinetic energy of the Earth core

(Appendix 3)

The core of the Earth drifts in the mantle in the direction counter to the Earth rotation. The angular velocity of the rotation of the Earth core is 0,2° less than the speed of the Earth rotation, i.e. the Earth core does not make a turn of 2π = 6,2831852 radian a day, but 6,2831757 radian a day. 0,2° a year = 0,0005749° a day = 0,0000095 radian a day.

The angular velocity of the Earth core rotation $\omega_я = \dfrac{6,2831757}{8,64 \times 10^4} = 7,272194 \times 10^{-5}$ rad/sec.

The angular velocity of the Earth rotation $\omega_з = \dfrac{2\pi}{8,64 \times 10^4} = 7,272205 \times 10^{-5}$ rad/sec.

Moment of inertia of the Earth $J_з = \dfrac{2}{5} m_з R^2_з = \dfrac{2}{5} \times 5,976 \times 10^{24} \times (6,371 \times 10^6)^2$ = 9,699 × 10³⁷ kg × m².

The moment of inertia of the core of the Earth $J_k = \dfrac{2}{5} m_k R^2_k = \dfrac{2}{5} \times 1,934 \times 10^{24} \times (3,471 \times 10^6)^2 = 9,32 \times 10^{36}$ кг × м²

The kinetic energy of the Earth $W_з = \dfrac{J\omega_з^2}{2} = \dfrac{9,699 \times 10^{37} \times (7,272205 \times 10^{-5})^2}{2} =$ 2,56 × 10²⁹ J.

Change of the kinetic energy of the Earth core $\Delta W_я = \dfrac{J(\omega_з^2 - \omega_я^2)}{2} =$

$\dfrac{9,32 \times 10^{36} [(7,272205 \times 10^{-5})^2 - (7,272194 \times 10^{-5})^2]}{2} = 7,456 \times 10^{22}$ J.

The total kinetic energy of the Earth equals to the sums of the energies of component parts, every second the Earth energy would lose 7,456 × 10²² J, even if the Earth were a completely hard body. But the fused mantle, which has high dynamic viscosity (coefficient of internal friction), absorbs the greater part of the energy of the core rotation. The difference of the angular velocities of the Earth and the core of the Earth is: $\Delta\omega = \omega_э - \omega_я$ = 7,272205 x 10^{-5} - 7,272194 x 10^{-5} = 1,1 x 10^{-10} radian/second.

The linear speed of the core rotation relative to the mantle on the border between the core and the mantle is V = $\Delta\omega$ x 3,471 x 10^6 = 3,818 x 10^{-4} m/sec = 33 m/day = 12 km/year.

Calculation of the thermal energy emitted by the Earth core
(Appendix 4)

Powerful thermal energy excreted at flow of the radial currents in the Earth core determines the heating of the core and the Earth on the whole.

According to Joule-Lenz law, the amount of the warmth excreted for time t at electric current flowing Q = I^2Rt = IUt = Pt.

Now we will determine the rating of the energy excreted by the Earth core at radial currents flowing.

As: P = UI = $\dfrac{U^2}{R}$; R = $\dfrac{\rho_{yd} l}{ds}$; R – resistance, Ohm;

ρ_{yd} – resistivity of a column with the length of 1m and section of 1 m^2, temperature considered, Ohm; ds - section of conductor, then Тогда P = $\dfrac{U^2 ds}{\rho_{yd} l}$;

U = VBl; l = r; V = $\dfrac{2\pi r}{T}$; U = $\dfrac{2\pi r^2 B}{T}$; P = $\dfrac{4\pi^2 r^3 B^2 ds}{T^2 \rho_{yd}}$. Function P (r) is distributed all over the volume of the core (ball). Since P = f(r), then dP = $\dfrac{12\pi^2 r^2 B^2 ds}{T^2 \rho_{yd}}$. Let's calculate the thermal energy excreted by the Earth core. In conformity [2] we will introduce coefficient 1/c. As U^2, then $1/c^2$.

P = $\iiint \dfrac{12\pi^2 B^2 r^2 ds}{c^2 T^2 \rho_{yd}}$ dx dy dz. In spherical coordinates:

P = $\iiint \dfrac{12\pi^2 B^2}{c^2 T^2 \rho_{yd}} \rho^2 \sin^2\theta \; \rho^2 \sin\theta \, d\theta \, d\varphi \, d\rho =$

$$\frac{12\pi^2 B^2}{c^2 T^2 \rho_{yo}} \int_0^{2\pi} d\varphi \int_0^R \rho^4 d\rho \int_0^\pi \sin^3\theta d\theta = \frac{12\pi^2 B^2}{c^2 T^2 \rho_{yo}} 6{,}28 \frac{R^5}{5} 1{,}333;$$

$$P = \frac{63{,}08 R^5 B^2}{c^2 T^2 \rho_{yo}}$$

At resistivity of a column with the length of 1m and section of $1m^2$: $\rho = 3{,}3 \times 10^{-6}$ Ohm, we calculate: the thermal energy excreted by the Earth core at radial electric currents flowing from the centre of the core to R = 3471 km, makes $8{,}2 \times 10^{10}$ watt/sec = $1{,}97 \times 10^{10}$ cal/sec = $6{,}92 \times 10^{17}$ cal/year.

I want morebooks!

Buy your books fast and straightforward online - at one of world's fastest growing online book stores! Environmentally sound due to Print-on-Demand technologies.

Buy your books online at
www.morebooks.shop

Kaufen Sie Ihre Bücher schnell und unkompliziert online – auf einer der am schnellsten wachsenden Buchhandelsplattformen weltweit! Dank Print-On-Demand umwelt- und ressourcenschonend produziert.

Bücher schneller online kaufen
www.morebooks.shop

KS OmniScriptum Publishing
Brivibas gatve 197
LV-1039 Riga, Latvia
Telefax: +371 686 204 55

info@omniscriptum.com
www.omniscriptum.com

www.ingramcontent.com/pod-product-compliance
Lightning Source LLC
Chambersburg PA
CBHW031542210526
45464CB00003B/1102